New Developments in Clinical Instrumentation

Editor

Leroy S. Hersh, Ph.D.
Staff Scientist
Biomedical Research Department
Research and Development Laboratories
Corning Glass Works
Corning, New York

CRC Press
Taylor & Francis Group
Boca Raton London New York

CRC Press is an imprint of the
Taylor & Francis Group, an **informa** business

First published 1981 by CRC Press
Taylor & Francis Group
6000 Broken Sound Parkway NW, Suite 300
Boca Raton, FL 33487-2742

Reissued 2018 by CRC Press

Library of Congress Cataloging in Publication Data

Main entry under title:

New developments in clinical instrumentation.

Bibliography: p.
Includes index.
1. Diagnosis, Laboratory — Data processing.
2. Medical laboratories — Automation. I. Hersh, Leory S.
RB38.N48 616.07'5'028 80-14505
ISBN 0-8493-5305-X

A Library of Congress record exists under LC control number: 80014505

Publisher's Note
The publisher has gone to great lengths to ensure the quality of this reprint but points out that some imperfections in the original copies may be apparent.

Disclaimer
The publisher has made every effort to trace copyright holders and welcomes correspondence from those they have been unable to contact.

ISBN 13: 978-1-315-89587-1 (hbk)
ISBN 13: 978-1-351-07497-1 (ebk)

Visit the Taylor & Francis Web site at http://www.taylorandfrancis.com and the
CRC Press Web site at http://www.crcpress.com

PREFACE

Quite understandably, there is a strong impetus to automate clinical laboratory procedures. Many clinical tests involve tedious, reptitious manipulations which must be performed with high reproducibility. Machines do not get bored, nor distracted (unless a power failure could be called a distraction). It was expected that relatively simple physical processes e.g. wet chemistries, testing for bacterial growth, could and would be automated.

A more recent aspect of automation has been made possible by the advent of the microprocessor or "computer-on-a-chip". The development of miniature, inexpensive micro-computers has resulted in the automation of relatively sophisticated processes.

The two aspects of automation are represented in this monograph. The automated control of physical processes, and automation of information processing. Not only does the automated instrument do away with the drudgery of boring manipulations, but it also "runs the machine", processes information, and makes "decisions".

There are elements of both aspects of automation in each of the chapters contained in this book. The areas discussed do emphasize more or less strongly either tight automated control of physical processes or the automation of information processing. They all represent the attempt of medical technology to yield more precise, accurate, less expensive, and faster results to aid in clinical diagnosis.

Leroy S. Hersh
March 1981

EDITOR

Leroy S. Hersh is a staff scientist at the Research and Development Laboratories of Corning Glass Works. He presently works in the Biomedical Research Department.

Dr. Hersh received his B. Ch. Eng. from Polytechnic Institute of Brooklyn and his Ph.D. in chemistry from the University of Chicago.

Dr. Hersh spent 2 years at the University of Chicago as a U.S. Public Health Service Post-doctoral fellow in the Department of Physiology. Since that time, he has been employed by Corning Glass Works for nearly 14 years.

Dr. Hersh is a past president of the Corning Sigma Xi Club. He is a member of the Biophysical Society, American Association for Clinical Chemistry, American Association for the Advancement of Science and Sigma Xi. He is also book review editor for the *Journal of Solid-Phase Biochemistry*.

CONTRIBUTORS

Donald P. Brezinski, Ph.D.
Senior Physicist
Applied Instrument Development Department
Corning Glass Works
Corning, New York

Charles N. Carter
Senior Project Engineer — Electrical
Corning Glass Works
Corning, New York

Stephen Carson Dunning
Senior Engineer
Amcor Incorporated
Raleigh, North Carolina

Karen Gooding
Research Associate
Department of Biochemistry
Purdue University
West Lafayette, Indiana

Laughton E. M. Miles, M.D.
Director
Laboratory for Chronobiological Research
Institute for Medical Research
San Jose, California
Clinical Assistant Professor of Medicine
Sleep Research Center
Stanford University Medical School
Stanford, California

W. Scott Ramsey, Ph.D.
Research and Development Laboratory
Corning Glass Works
Corning, New York

Fred E. Regnier, Ph.D.
Professor of Biochemistry
Biochemistry Department
Purdue University
Lafayette, Indiana

Charles H. Rogers, Ph.D.
General Products Manager
Corning Medical and Scientific
Corning Glass Works
Corning, New York

TABLE OF CONTENTS

Scanning Fluorometers and Densitometers: Principles and Performance Analysis 1
D. P. Brezinski

Automated Leukocyte Differential Analyzers 35
Charles N. Carter

Computers in Clinical Instrumentation 57
Stephen C. Dunning

The Use of High Performance Liquid Chromatography in the Clinical Laboratory 71
Karen M. Gooding and Fred E. Regnier

Automated Microbiology 103
W. Scott Ramsey

Automation of Radioimmunoassay...127
Charles H. Rogers and Laughton E. M. Miles

Index 147

SCANNING FLUOROMETERS AND DENSITOMETERS: PRINCIPLES AND PERFORMANCE ANALYSIS

D. P. Brezinski

TABLE OF CONTENTS

I. Abbreviations 2

II. Introduction 2

III. Densitometric Theory 4

IV. Densitometric Instrumentation and Errors 5
- A. Instrument Configurations 5
- B. Densitometric Errors 6
  1. Light-Source/Detector Stability 6
  2. Errors due to Inappropriate Slit Size 7
  3. Fixed Offset Errors (Electronic, Optical) 12
  4. Bandwidth Error 13
  5. Wavelength Selection 16

V. Fluorometric Theory 16
- A. Fluorescence Activity 18

VI. Fluorometric Instrumentation and Errors 20

VII. Instrumentation and Performance Specification 21
- A. System Description 21
- B. Performance Specifications 24
  1. Range 24
  2. Linearity 25
  3. Pen Recorder Accuracy 25
  4. Integration Accuracy 26
  5. Electronic Integrator Accuracy 26
  6. Auto-Zero and Auto-Gain Accuracies 27
  7. Effective Optical Slit and Off-Image Blocking 27
  8. Optical/Electronic Rise Time 28

VIII. Performance Testing and Quality Control 29
- A. Linearity, Linear Range 29
  1. Densitometric Linearity 29
  2. Fluorometric Linearity 31
- B. Auto-Zero, Auto-Gain, Pen-Recorder, and Integrator Accuracies 32
- C. Scan Rate Uniformity 32
- D. Slit Response Profile 32

IX. Conclusion 33

References 34

I. ABBREVIATIONS

a = absorptivity
A = absorbance
A_o = peak absorbance of fraction
AR = absorbance ratio = (average absorbance in slit)/(peak absorbance)
c = sample concentration
F = fluorescence emission
FA = fluorescence activity
I_o = incident light intensity
I = transmitted light intensity
ℓ = optical path length through sample
L = linearity factor = A_{resp}/A_{ave}, where
  A_{resp} = densitometric response to sample
  A_{ave} = average absorbance of sample across instrument bandpass or slit
OD = optical density (unit)
QE = quantitation efficiency = measured value/true value of amplitude or area
RSW = relative slit width = half-power slit width/half-amplitude peak width
T = transmittance
δ = relative variation in absorbance
Δ = absorbance error
ε = source intensity fluctuation, offset error
λ = wavelength
ϕ = fluorescence quantum efficiency

II. INTRODUCTION

A scanning fluorometer/densitometer is an instrument for measuring the spatial distribution of fluorescent or optically absorbant chemical species in a sample. Such instruments are routinely used in the clinical laboratory for quantitating electrophoretic separations of serum proteins, cardiac isoenzymes, hemoglobins, lipoproteins, etc. Electrophoretic separations are treated with specific stains or reagents that render the desired species visible as colored or fluorescent bands. While the results are often evaluated by visual inspection, the use of instrumentation for more precise quantitation has become increasingly popular. Quantitative output is usually in the form of a recorder tracing showing sample absorbance or fluorescence as a function of position along the separation. This output is important for the evaluation of patterns that are interpreted largely in terms of band shape, amplitude, or position. In other tests, such as lactate dehydrogenase and creatine kinase cardiac isoenzymes, band shape and position are largely invariant, and the relative quantity of the different fractions is of diagnostic significance. Here the clinician is interested in obtaining numerical values for the amount of activity in the different bands, in either absolute or relative terms. This is achieved by integrating the area under the different bands and representing each band as a fraction of the total area.

Scanning densitometers and fluorometers for clinical applications are marketed by a number of companies, including Beckman, Corning, Gelman, Helena, and Transidyne General. Most of these companies are also major distributors of electrophoresis equipment and supplies. A feature-by-feature comparison of the instruments offered by various manufacturers is beyond the scope and intent of this article. A recent article by Alpert[1] gives a limited functional appraisal of some of these instruments.

The primary motivation of the present paper is the lack of adequate information on the performance characteristics of available clinical fluorometer/densitometers. Both

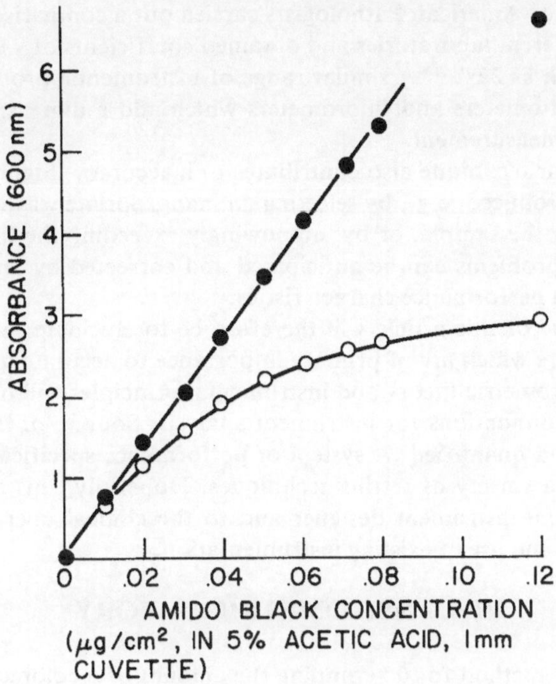

FIGURE 1. Comparative linearity of two clinical scanning densitometers (amido black dye at 600 nm).

manufacturers and users tend to evaluate their instruments in terms of design rather than performance. An instrument is usually described in terms of convenience features, sample capability, and design details, such as type of photodetector, light source, etc. These factors are of course important to the user, but little is said about an equally important factor: namely, how well the instruments actually perform in an analytical sense. Manufacturers' specifications (e.g., ''linear range: 0 to 2 OD'') are often meaningless as indicators of performance, since error tolerances, measurement conditions, and type of sample are rarely specified.

A lack of attention to instrument performance characteristics as opposed to features would be understandable if instrumental errors were always negligible compared to the variability in electrophoretic results due to the chemistries. I have observed, however, that different instruments sometimes give quite different results for the same sample, indicating that instrumental factors are at times significant. For example, in direct clinical comparisons with identical serum protein samples, we have observed mean differences as large as 7% (absolute) in the albumin percentages reported by different densitometers. These discrepancies appeared attributable to differences in instrument linearity with dyes (see Figure 1).

Understandably, accuracy tends to be a secondary concern of the less sophisticated instrument user, who may be only vaguely aware of the distinction between accuracy and precision and feels no reason to doubt his results if they are reproducible. A contributory factor is the lack of simple but adequate procedures for independently testing instrument accuracy. Consistent instrument errors may easily pass unnoticed in a clinical setting, since they can be absorbed into the normal values established for a procedure. Problems may become apparent only when results are exchanged between laboratories or instrumentation is changed. Accuracy and precision problems with ''cuvette''-type spectrophotometers have been amply documented.[2-4,19] For example, in 1973 the College of American Pathologists carried out a comparison of spectrophotometers in 132 different laboratories and obtained coefficients of variation in absorbance ranging as high as 22%.[13] A similar range of instrumental problems seems likely with scanning densitometers and fluorometers which add a dimension of complexity to the photometric measurement.

Improper operator technique also contributes to inaccuracy. Significant densitometric error can be introduced, e.g., by selecting an inappropriate wavelength or slit size, improperly aligning the sample, or by unknowingly exceeding the linear range of the instrument. Many problems can be anticipated and corrected by an operator who is aware of instrument performance characteristics.

The main purpose of this article will therefore be to elucidate the instrument performance parameters which are of primary importance to accurate quantitation. Densitometric and fluorometric theory and instrumental principles will be discussed to the extent required as foundations for instrument analysis. Sources of instrumental error will be identified and quantified. A system of performance specifications will be suggested, along with a variety of testing techniques. Hopefully, this discussion will be valuable, both to the instrument designer and to the clinical operator interested in accurate evaluation and use of existing instrumentation.

III. DENSITOMETRIC THEORY

Densitometry is a method for determining the amount of a colored chemical species in a sample by measuring the amount of light it absorbs at a certain wavelength. In a densitometric measurement, a beam of light of known intensity and wavelength is passed through the sample. A portion of this light is absorbed by the colored species. The remaining portion of light passes on to a detector, where it is measured and converted to a signal related to the species concentration.

The fundamental equation of densitometry is the Beer-Lambert law:

$$I/I_o = 10^{-ac\ell}$$

where I_o is the intensity of light entering the sample, I is the intensity of light leaving the sample, a is a wavelength-dependent constant known as the *absorptivity*, c is the sample concentration, and ℓ is the sample pathlength, i.e., the distance that light travels through the sample. c is usually expressed in moles per liter, and ℓ is usually expressed in centimeters. In this case a is known as the *molar absorptivity* and has unit dimensions of liter-centimeter per mole.

In essence the Beer-Lambert law states that the fraction of light passed by the sample falls off exponentially with the amount of light-absorbing species. This law generally holds in the absence of special effects such as dye molecule dimerization. The ratio $T = I/I_o$ is called the *transmittance*. The quantity $A = -\log T = \log I_o - \log I$ is known as the *absorbance*. If the Beer-Lambert law holds, we have $A = -\log(10^{-ac\ell}) = ac\ell$. This allows the area concentration of absorbing species, $c\ell$, to be determined from the measured absorbance A and known absorptivity a.

The value of the absorptivity a generally varies greatly with wavelength. A *spectrophotometer* is used to determine a dye "spectrum" (i.e., a as a function of wavelength) by measuring absorbance as a function of wavelength for a sample of known concentration and pathlength.

A *scanning densitometer* is used to determine the spatial distribution of dye over a sample such as an electrophoresis plate. For this application, a fixed wavelength is selected. A light beam defined by a densitometric slit is scanned over the sample, and

Table 1
COMPARISON OF SEMISPECULAR AND DIFFUSE DENSITIES OF PHOTOGRAPHIC STEP TABLET (NBS DIFFUSE DENSITY STANDARD 67101)

Step	NBS density (diffuse)	Measured density (semispecular, θ<13°)	Density ratio: semispecular/ diffuse
1	0.06	0.09	1.5
2	0.20	0.315	1.58
4	0.495	0.752	1.52
7	0.95	1.38	1.45
11	1.55	2.17	1.40
14	2.005	2.73	1.36
17	2.45	3.30	1.34
21	3.045	4.02	1.32

absorbance as a function of position is determined. Since absorbance is proportional to $c\ell$, this gives the distribution of dye over the sample. Often a relative rather than absolute measurement of dye distribution is all that is required.

Use of terminology in this field is somewhat variable. The favored term for the measurement of absorption is *absorptiometry.* If light scattering, as well as absorption, contributes significantly to the attenuation of the incident beam, the measurement is often called *densitometry,* and a different set of terms is sometimes employed. The ratio $T = I/I_o$ is in this case known as the *transmission,* $D = -\log T$ is known as the *optical density,* and $\varepsilon = D/c\ell$ is known as the *extinction coefficient.* Though absorbance or optical density is a dimensionless number, it is often expressed in "OD" units. As discussed below, many clinical instruments measure optical density rather than absorbance, so the terms "densitometer" and "densitometry" are appropriate.

IV. DENSITOMETRIC INSTRUMENTATION AND ERRORS

A. Instrument Configurations

In a measurement of true absorbance, light scattered but not absorbed by the sample is counted as part of the transmitted beam. This imposes a severe constraint on optical design, since the optics must be capable of collecting light scattered at extreme angles and routing it to the detector. This requires a large optical aperture (slit or lens close to sample) and makes it harder to sharply define the area of sample being quantitated.

In a measurement of optical density, light scattered by more than a certain angle is not detected. If the acceptance angle is less than 1°, the measured quantity is termed *specular density,* while if the acceptance angle approaches 90°, the measured quantity is termed *diffuse density.* Intermediate measurements are called "semispecular" and are typical of clinical densitometers. These distinctions are of interest to the clinical user, since photographic step tablets are often used to check instrument linearity. Such tablets (e.g., National Bureau of Standards Diffuse Density Step Tablet) scatter light strongly and are generally calibrated in units of diffuse optical density. A properly functioning instrument measuring specular or semispecular density will appear to function nonlinearly when tested with such a tablet, even though its performance is linear with nonscattering samples such as Wratten neutral density filters or dyes (see Table 1).

The choice between specular and diffuse instrumentation is not clear cut. A diffuse density measurement is less subject to uneven background due to light scattering from poorly cleared cellulose acetate samples or lipemic sera, for example. On the other hand, specular optics impose fewer design constraints and may offer higher resolution. For most clinical samples, the difference between specular and diffuse density will be negligible.

"Dual-Beam" densitometry is often used in applications where I_o is apt to vary. In a dual-beam measurement, the light beam is alternately routed between a sample and reference path. The reference path may include a "blank" sample. I is measured over the sample path, and I_o is measured over the reference path. While this methodology can be used to compensate for short-term fluctuations in the value of I_o, its main application is in spectrophotometry, where the effective value of I_o changes significantly with wavelength due to changes in light source output, detector sensitivity, and absorption by the solvent. In single-beam densitometry, both I_o and I are measured over the same optical path. The absorbance reading is manually or automatically zeroed for the value of I_o corresponding to the blank sample or scan baseline. If the electronics and light source are stable, I_o does not change, and an accurate measurement can be made. The single-beam methodology is generally employed in clinical scanning densitometers.

Logarithmic conversion of the transmittance signal can be accomplished in several ways. The transmittance signal can be directly converted into an absorbance signal by a logarithmic electronic amplifier. Alternately, the transmittance signal can be nulled by a second signal which is exponentially related to displacement of the chart recorder pen. For example, an exponentially wound slide wire can be used as the chart recorder position sensor. Similarly, displacement of the recorder pen can be coupled to insertion of an optical density wedge into a reference beam. Sample density is then given by the wedge displacement required to give equal reference and sample signals.

B. Densitometric Errors

Densitometric measurements are commonly made in the absorbance range of 0.01 to 3. Outside of these limits, accurate measurements are more difficult to make due to various sources of error. The following is a discussion of the performance parameters which are crucial to accurate densitometry over an extended range.

1. Light-Source/Detector Stability

At very low absorbance, the quantities log I and log I_o are large and nearly equal, making accurate determination of the difference difficult. Also, small fluctuations of I_o due to source or detector instability become significant. This error can be quantitated as follows. Let ε be the amount that the incident light beam deviates from its initially measured value, I_o. The absorbance deviation or error, Δ, is then given by

$$
\begin{aligned}
\Delta &= (\text{calculated absorbance}) - (\text{true absorbance}) \\
&= \log\left(\frac{I_o}{I}\right) - \log\left(\frac{I_o + \epsilon}{I}\right) \\
&= -\log\left(1 + \frac{\epsilon}{I_o}\right) \\
&\approx -\frac{1}{2.303}\,\frac{\epsilon}{I_o} \quad \text{if } \frac{\epsilon}{I_o} \ll 1
\end{aligned}
$$

For example, this equation indicates that absorbance measurement to an accuracy of $\Delta = 0.001$ requires a source/detector stability of 0.23% ($\varepsilon/I_o \approx 0.0023$). A value of Δ <0.0002 is attainable with an incandescent tungsten filament lamp and a PIN silicon photodiode detector, indicating that inexpensive single-beam instrumentation can be quite adequate for clinical scanning of low-density samples. In practice, quantitation

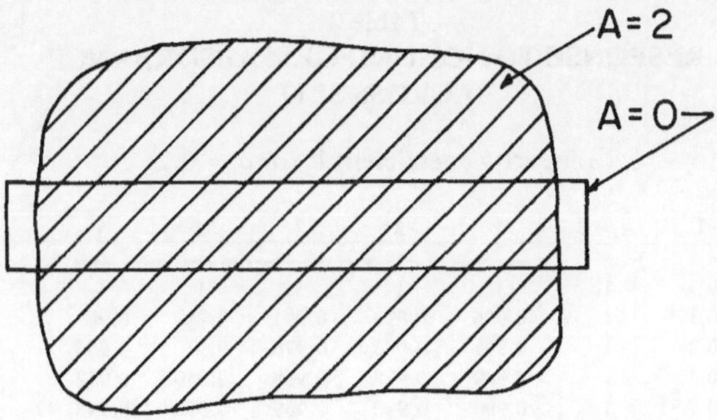

FIGURE 2. Densitometric error due to excessively long slit. Slit width 10% wider than sample. Average transmittance = 0.1 × 1 + 0.9 × 10^{-2} = 0.109. Measured absorbance = log (0.109) = 0.96. Average (true) absorbance = 0.1 × 0 + 0.9 × 2 = 1.8. Error = (0.96 − 1.8)/1.8 = −47%.

at low densities is usually limited by background "noise" in the sample from scratches in the electrophoresis film, uneven film clearing, etc.

At very high absorbance, measurements become extremely sensitive to very small instrumental errors. This error sensitivity is due to the fact that the electronic signal at high absorbance is itself extremely small. For instance, if the absorbance baseline gives a signal of 1 V, the electronic signal at 5 OD will be five orders of magnitude lower or 10 μV. An accuracy of 1% at 5 OD requires an accuracy to within 1 millionth of the baseline signal level. Measurement errors of this magnitude arise from a number of sources and are extremely difficult to avoid. These sources include slight offsets in the electronic amplifiers, slit/sample misalignment and incompatibility, stray light entering the detector from ambient room light or spurious light sources within the instrument, and inadequate spectral filtering of the densitometric light source. These error sources manifest themselves as densitometric nonlinearity; i.e., a plot of measured absorbance vs. sample concentration departs from a straight line at high values of absorbance (see Figure 1).

Most causes of linearity problems at high absorbance can be understood in terms of the following principle: for accurate absorbance measurement, the absorbance must be the same for all optical paths and wavelengths contributing to the measured transmittance. This principle is illustrated in the following error analyses.

2. Errors due to Inappropriate Slit Size

For linear results, the densitometric slit must be small enough that sample concentration does not vary significantly across the slit at any point in the scan. For example, severe error is introduced if the slit is too long for the sample bands (see Figure 2). This error arises because averaging the transmittances corresponding to different absorbance levels and taking the logarithm does not give the correct value for the average absorbance. In practice, the densitometric slit should be somewhat shorter than the band of sample applied to the electrophoresis track. This is in distinction to fluorometry, where a slit long enough to sweep the entire width of the track can be used. Note that since the full length of electrophoresis bands cannot be accurately scanned in densitometry, accuracy depends on scanning a representative portion of each band. Therefore, erroneous fraction values may result if band length varies excessively from fraction to fraction or if the scan path is poorly aligned with the track centerline. To avoid errors due to inconsistent fraction geometry, serum samples should be applied

Table 2
RESPONSE TO NONUNIFORM ABSORBANCE ACROSS SLIT

Linearity coefficient L for $\delta = 0.2$

f	$A_o = 0$	1	2	3	4	5
0	1	1	1	1	1	1
0.1	1	0.996	0.993	0.991	0.989	0.987
0.3	1	0.991	0.983	0.976	0.970	0.966
0.5	1	0.989	0.978	0.968	0.959	0.952
0.7	1	0.990	0.980	0.969	0.958	0.948
0.9	1	0.996	0.990	0.984	0.976	0.967
1	1	1	1	1	1	1

Note: Fractional areas of slit

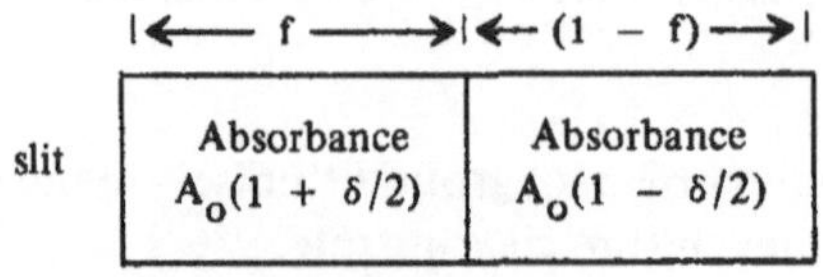

$$L = \frac{A_{resp}}{A_{ave}} = \frac{-\log\left[\int_0^f 10^{-(1+\delta/2)A_o}dx + \int_f^1 10^{-(1-\delta/2)A_o}dx\right]}{\int_0^f (1+\delta/2)A_o dx + \int_f^1 (1-\delta/2)A_o dx} =$$

$$= \frac{A_o - \log\left[10^{-\delta A/2}f + 10^{\delta A/2}(1-f)\right]}{A_o\left[1 + \delta f + \delta/2\right]}$$

uniformly across the sample wells, and overloading should be avoided. During scanning, the sample should be positioned so that the densitometric slit passes straight down the centerline of the track.

It is well known that full resolution of closely spaced electrophoretic peaks requires an adequately narrow and sharply focused slit. It is not widely appreciated, however, that a wide or poorly focused slit also causes linearity errors at high absorbance. This is due to the fact that a high-absorbance peak will be significantly depressed by slight amounts of light entering the slit from adjacent, less-absorbent areas of the profile. In fluorometric mode, poor resolution merely smears out the pattern without affecting linearity.

Linearity error due to sample absorbance variation across the slit can be quantitatively evaluated by comparing instrument response with the average absorbance in the slit. Table 2 shows the ratio of instrument response to average absorbance for different distributions of absorbance across the slit. The results indicate that if sample absorbance changes less than ±10% across the slit, linearity error is less than ~1% per OD. This provides a crude guideline for selecting slit dimensions that will give accurate quantitation while scanning over density transitions in the sample.

This guideline is, however, hard to apply to the actual scanning of electrophoretic peaks, where the absorbance profile across the slit constantly changes. Greater insight is provided by calculating the efficiency in quantitating the area of an absorbance peak by scanning with a slit of given relative width. Quantitation efficiency as a function of maximum absorbance and relative width of slit and peak is given in Table 3 and

Table 3
AREA QUANTITATION EFFICIENCY*

RSW	QE_{AREA} $A_o = 1$	2	3	4	5
	Rectangular Slit Sensitivity Profile				
0.1	0.998	0.996	0.994	0.993	0.991
0.2	0.993	0.985	0.978	0.971	0.965
0.3	0.984	0.968	0.953	0.939	0.926
0.4	0.971	0.944	0.920	0.898	0.880
0.5	0.951	0.917	0.882	0.852	0.828
0.6	0.940	0.886	0.840	0.803	0.773
0.7	0.921	0.853	0.797	0.753	0.719
0.8	0.902	0.818	0.753	0.704	0.666
0.9	0.881	0.784	0.711	0.656	0.615
1.0	0.861	0.750	0.670	0.611	0.568
	Triangular Slit Sensitivity Profile				
0.1	0.996	0.993	0.989	0.985	0.982
0.2	0.985	0.971	0.957	0.943	0.931
0.3	0.968	0.937	0.908	0.881	0.857
0.4	0.946	0.895	0.848	0.807	0.771
0.5	0.920	0.847	0.783	0.729	0.684
0.6	0.893	0.797	0.717	0.654	0.603
0.7	0.865	0.749	0.657	0.587	0.534
0.8	0.837	0.704	0.604	0.531	0.477
0.9	0.812	0.665	0.559	0.485	0.431
1.0	0.788	0.630	0.522	0.448	0.394

* QE = Measured area/true area for Gaussian absorbance peak scanned by moving slit.

plotted in Figure 3. The absorbance band is presumed to be Gaussian in shape (which is appropriate for a diffusing species), and the spatial sensitivity (power) profile of the slit is presumed to be rectangular (which is appropriate for an idealized masked band-pass filter) or triangular (which is appropriate for most grating or prism monochrometers with equal entry and exit slits). Actual slit configurations will usually lie between these two extremes. Peak and slit widths are measured at half-amplitude and power, respectively. The plotted quantitation efficiency is the ratio of measured integrated absorbance (area) to true integrated absorbance. The plots indicate a drastic drop in quantitation with increasing relative slit width (RSW = slit width/peak width). For example a square slit with 0.3 RSW gives an error of −5% for a 3-OD peak. However, in most clinical determinations, absolute area is of no interest, and only the relative areas of different fractions are desired. If one approximates the error curves with straight lines, one finds that the relative error for a two-peak profile reaches a very broad maximum at approximately a 3-to-1 peak ratio (valid for RSW <~0.8) and that for this ratio the error as a percent of total area depends only on quantitation efficiency (QE) and not explicitly on total absorbance. Fraction error levels corresponding to different QEs of the larger peak are in Table 4 and in Figure 3. For example, with a large peak at 3 OD and small peak at 1 OD (75%/25% area ratio) and with an RSW of 0.3, fraction error is about 0.6%. Thus, the results obtained would be 74.4 and 25.6%. The error is much less than would be expected from the quantitation efficiency loss of the larger peak (−5%). The error would be even less with a

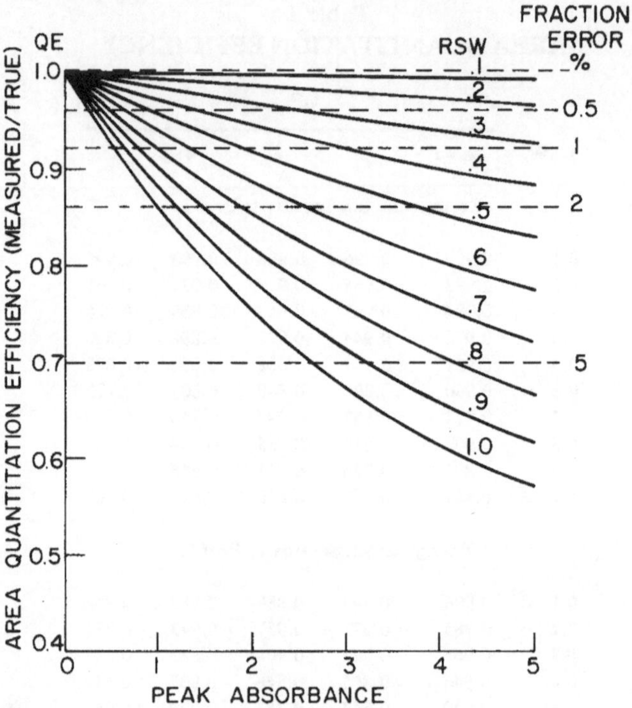

FIGURE 3. Quantitation efficiency and maximum two-peak fraction errors for rectangular slit profiles (from Tables 3 and 4).

Table 4
FRACTION ERROR FOR TWO-COMPONENT PROFILE WITH LINEARLY DECREASING QUANTITATION EFFICIENCY

Quantitation efficiency of larger peak	Fraction ratio for maximum error	Maximum fraction error[a] (percent of total)
1.00	—	0
0.99	3.00	0.13
0.98	3.00	0.25
0.95	3.00	0.65
0.9	3.01	1.36
0.8	3.03	3.00
0.7	3.04	5.00
0.6	3.08	7.50
0.5	3.11	10.72

[a] Very broad maximum.

different area ratio. As a general conclusion, an RSW of 0.3 should be adequate for most clinical determinations.

If the sample fractions are quantitated on the basis of amplitude rather than area, then quantitation accuracy depends only on the peak-centered response. Amplitude quantitation efficiency as a function of maximum absorbance and RSW is in Table 5. The ratio, AR, of the average absorbance in the slit to the peak absorbance is also tabulated. AR is independent of peak absorbance. Overall response to a peak is given by the product of the AR and QE factors. For example, if a rectangular, 0.8 RSW slit is used to quantitate a 3-OD peak, the average absorbance in the slit is $A_o \times AR = 3 \times$

Table 5
AMPLITUDE QUANTITATION EFFICIENCY[a]

RSW	AR	QE$_{Amplitude}$ $A_o = 1$	2	3	4	5
Rectangular Slit Sensitivity Profile						
0.1	0.998	1.000	1.000	1.000	1.000	1.000
0.2	0.991	1.000	1.000	1.000	1.000	1.000
0.3	0.980	1.000	0.999	0.999	0.998	0.998
0.4	0.964	0.999	0.998	0.996	0.995	0.994
0.5	0.945	0.997	0.994	0.991	0.988	0.984
0.6	0.923	0.994	0.988	0.982	0.976	0.970
0.7	0.897	0.990	0.979	0.968	0.958	0.947
0.8	0.870	0.983	0.966	0.948	0.932	0.918
0.9	0.841	0.975	0.949	0.924	0.901	0.882
1.0	0.810	0.964	0.928	0.894	0.865	0.840
Triangular Slit Sensitivity Profile						
0.1	0.995	1.000	1.000	1.000	1.000	1.000
0.2	0.982	0.999	0.999	0.998	0.998	0.997
0.3	0.960	0.997	0.995	0.991	0.988	0.984
0.4	0.932	0.992	0.983	0.973	0.962	0.950
0.5	0.899	0.982	0.961	0.938	0.913	0.887
0.6	0.862	0.967	0.928	0.885	0.843	0.805
0.7	0.823	0.946	0.884	0.821	0.764	0.717
0.8	0.783	0.923	0.835	0.754	0.687	0.634
0.9	0.744	0.896	0.785	0.690	0.617	0.561
1.0	0.707	0.869	0.738	0.633	0.557	0.500

Note: QE = Measured absorbance/average absorbance for slit centered on Gaussian absorbance peak.

[a] RSW = relative slit width; AR = absorbance ratio = average absorbance in slit/peak absorbance; A_o = peak absorbance.

0.870 = 2.61, but because of linearity error, response will be further reduced by the factor QE = 0.948 to give a reading of 2.47 OD. AR decreases very rapidly with slit width, indicating that a quite narrow slit is required to yield accurate peak absorbance values, even at low optical density (e.g., RSW 0.3 is required for 2% accuracy with a rectangular slit). On the other hand, if linearity rather than accuracy is the requirement, a considerably wider slit can be used. For example, a 0.7 RSW rectangular slit will give ±2% linearity up to 4 OD. For low RSWs, the QE vs. absorbance curves for amplitude quantitation are relatively straight lines, similar to those in Figure 3. Thus, the quantitation of relative amplitudes gives the same relationship between quantitation efficiency and fraction error as in Table 3. Thus, a rectangular, 0.9 RSW slit quantitating 3-OD peak has a QE of 0.92, but yields a fraction error of only 1% in determination of a 3-OD/1-OD peak ratio.

Finally, a comparison of Tables 3 and 5 reveals that amplitude quantitation is markedly more linear than area quantitation. Area quantitation yields nearly tenfold higher linearity error with the intermediate-RSW slits. On the other hand, amplitude quantitation is affected by variation in fraction width, whereas area quantitation is not. When a sample fraction is selectively broadened by diffusion or tailing, maximum concentration is decreased, but total material is conserved. Accurate amplitude quantitation therefore requires the half-amplitude width of all fractions to be similar; area quantitation does not. Which quantitation technique is preferred, area or amplitude, will depend on the uniformity of the bands to be quantitated. For example, amplitude quantitation can be expected to give good results with electrophoretic LDH isoenzyme profiles, but very poor results with serum proteins. In the face of uncertainty, area quantitation with a quite narrow slit is probably the safest choice.

Quantitation of irregularly shaped spots is greatly improved by zig-zag and array detection methods which in effect scan laterally as well as longitudinally along the sample track. In zig-zag scanning, a very small detection aperture is swept back and forth transversely to the scan direction. The results is a zig-zag scan path which separately and accurately quantitates all portions of the track. Each fraction is resolved onto a series of cross-sectional profiles, the areas of which are added to give total material. The same result can be achieved by using a linear, side-by-side array of small detectors to scan the track. The individual outputs are logarithmically processed and then added together to give total absorbance across the width of the track. These scanning methods also allow automatic compensation for longitudinal variations in background, since the absorbance level at the edges of the track can be instrumentally subtracted from absorbances in the center portion of the track containing the sample. For example, thin-layer chromatography (TLC) solvent-front artifacts can be eliminated in this way.

Dual wavelength densitometry is another background-reducing feature available on some instruments. In this technique, a chopped light beam at two different wavelengths is used to scan the sample, and the difference in absorbance at the two wavelengths is outputted. Since $A_1 - A_2 = (a_1 - a_2)\, cl$, the output is still linear in c, provided that absorptivities a_1 and a_2 are different. If wavelengths are chosen which maximize $a_1 - a_2$ for the species of interest and minimize it for interfering species, specific interferences can be eliminated. Also this technique tends to eliminate artifacts due to light scattering, since scattering extinctions generally do not change rapidly with wavelength. For example, the procedure is effective in reducing background artifacts due to thickness variations in TLC plates.

3. Fixed Offset Errors (Electronic, Optical)

Fixed offset errors in the measured transmittance cause large linearity errors at high absorbance levels. Electronic offsets arise from sources such as dark current in the photodetector and offset voltages in the amplifiers preceding the logarithmic stage. Optical offsets arise from stray light. This is light from within the instrument or from the external environment which enters the detector without first passing through the sample. Fixed optical offsets may also occur because of inadequate spectral filtering of the densitometric light beam. For example, Ponceau S is generally quantitated near its absorbance maximum of 520 nm. If the 520-nm beam is contaminated with a 0.01% content of 600-nm light, which Ponceau does not absorb, this introduces an error signal which is as large as the true signal at absorbance 4.

Let ε be the offset error in measured intensity. Define $T_\varepsilon \equiv \varepsilon/I_o$ as the "transmittance" error. Then the absorbance error, Δ, is given by the difference between measured and true absorbance, or

$$\Delta = \log\left(\frac{I_o + \epsilon}{I + \epsilon}\right) - \log\left(\frac{I_o}{I}\right)$$

$$= \log\left(\frac{1 + T_\epsilon}{1 + T_\epsilon/T}\right)$$

$$\approx -\log(1 + T_\epsilon/T)$$

If $T_c > 0$, densitometric response falls off rather abruptly above the $T \approx T_c$ absorbance level. T_c may be less than zero with electronic offsets, in which case measured absorbance rises towards infinity as T approaches T_c.

In practice, total offset should be kept at least an order of magnitude below transmittance at the highest absorbance to be measured.

4. Bandwidth Error

Densitometric response is based upon the overall attenuation of the beam of light passed through the sample. If this light contains more than one wavelength component and if sample absorptivity is not the same at the different wavelengths, then deviations from linearity will occur at high absorbance levels. Error of this sort is introduced if sample absorbance varies excessively over the bandpass (wavelength spread) of the instrument's monochrometer or bandpass filter. The theory of spectral bandpass errors has also been treated by Hardy and Young,[9] Eberhardt,[10] and Broderson.[11]

Let $\varepsilon(\lambda)$ be sample absorptivity as a function of wavelength. For a given monochrometer or filter setting, let $W(\lambda)$ designate the product of incident light intensity and detector sensitivity at wavelength λ. $W(\lambda)$ is in effect the instrumental "weight" given to the wavelength λ. Instrumental densitometric response as a function of concentration Cl is then given by:

$$A_{resp} = -\log \left(\int W(\lambda)\, 10^{-\epsilon(\lambda)Cl} d\lambda \Big/ \int W(\lambda)\, d\lambda\right)$$

The weighted-average absorbance of the sample over the instrument bandpass is given by:

$$A_{ave} = Cl \int W(\lambda)\, \epsilon(\lambda)\, d\lambda \Big/ \int W(\lambda)\, d\lambda$$

A_{ave} is linear with sample concentration. It is instructive to evaluate the ratio of A_{resp} to A_{ave} for two idealized instrument bandpass characteristics. Multicoated interference filters are often designed to have a roughly square wavelength bandpass, describable by:

$$W(\lambda) = \begin{cases} W_0 \text{ for } |\lambda - \lambda_0| \leq \tfrac{1}{2} B \\ 0 \text{ for } |\lambda - \lambda_0| > \tfrac{1}{2} B \end{cases}$$

where λ_0 is the center, and B is the width of the bandpass. On the other hand, a prism or grating monochrometer with equal entrance and exit slits has an approximately triangular wavelength bandpass,[16] describable by:

$$W(\lambda) = \begin{cases} W_0(1 - |\lambda - \lambda_0|/B) & \text{for } |\lambda - \lambda_0| \leq B \\ 0 & \text{for } |\lambda - \lambda_0| > B \end{cases}$$

where B is the half-intensity bandwidth.

Clinical samples are usually scanned with the monochrometer centered on an absorbance peak. However, if the sample is excessively absorbent, it may be advantageous to scan it "off-peak", i.e., with the monochrometer centered to one side or other of the absorbance maximum. A peak-centered measurement can be crudely ap-

Table 6
LINEARITY FACTOR L = $A_{response}/A_{average}$ AS A FUNCTION OF CENTRAL ABSORBANCE AND RELATIVE ABSORBANCE VARIATION OVER BANDPASS

		A_o[a]					A_o[a]				
δ[a]	RSW	1	2	3	4	5	1	2	3	4	5
		Square bandpass; resonance peak					**Square bandpass; off-peak slope**				
0.05	0.23	1.000	0.999	0.999	0.999	0.999	1.000	1.000	0.999	0.999	0.999
0.1	0.33	0.999	0.998	0.997	0.995	0.994	0.999	0.998	0.997	0.996	0.995
0.2	0.50	0.995	0.990	0.985	0.980	0.975	0.996	0.992	0.989	0.985	0.982
0.3	0.65	0.988	0.976	0.964	0.953	0.942	0.991	0.983	0.975	0.967	0.960
0.4	0.82	0.978	0.955	0.934	0.914	0.897	0.985	0.970	0.957	0.944	0.934
0.5	1.00	0.962	0.925	0.893	0.865	0.843	0.976	0.954	0.934	0.917	0.902
		Triangular bandpass; resonance peak					**Triangular bandpass; off-peak slope**				
0.05	0.23	0.998	0.997	0.995	0.993	0.990	1.000	0.999	0.999	0.998	0.998
0.1	0.33	0.994	0.987	0.979	0.971	0.962	0.998	0.996	0.994	0.992	0.991
0.2	0.50	0.978	0.954	0.927	0.901	0.876	0.992	0.985	0.977	0.970	0.963
0.3	0.65	0.957	0.910	0.864	0.822	0.787	0.983	0.966	0.950	0.935	0.921
0.4	0.82	0.932	0.862	0.800	0.750	0.709	0.970	0.940	0.913	0.889	0.867
0.5	1.00	0.904	0.814	0.741	0.685	0.642	0.953	0.909	0.868	0.834	0.805

[a] A_o = sample absorbance at center of bandpass; δ = fractional change in absorbance over half-power bandpass (= $(A_{max} - A_{min})/A_o$); RSW = slit half-amplitude bandwidth/dye half-amplitude bandwidth

proximated by assuming a resonance-shaped $(1 + k(\lambda - \lambda_o)^2)^{-1}$ absorbance distribution centered about the measurement bandpass, while an off-peak measurement can be approximated by a linearly changing absorbance across the measurement bandpass.

Values for the ratio A_{resp}/A_{ave} for different sample and bandpass configurations are in Table 6 as a function of absorbance and a factor, δ, representing fractional absorbance change across the bandpass. In all cases, A_{resp}/A_{ave} approaches one at very low absorbance levels, indicating that instrument response approaches the value of the weighted-average absorbance. As absorbance increases, this ratio falls below one at a rate which depends on absorbance variation across the bandpass. The decrease in the ratio A_{resp}/A_{ave} indicates the fractional deviation from linear response. Table 6 also includes RSW values corresponding to each δ, based on a resonance-shaped peak. While spectral error data is generally tabulated as a function of RSW, in practice the δ value is easier to determine from a measured spectrum and is less sensitive to band shape. Error curves plotted from Table 6 data yield fairly straight lines (see Figure 4), so the fraction error values of Table 4 again apply to relative measurements.

An examination of Table 6 and Figure 4 leads to some interesting conclusions. Under some conditions, a quite large absorbance variation can occur across the bandpass without introducing excessive linearity error. A 30% absorbance change (δ = 0.3) over a square bandpass introduces only about 1% deviation from linearity per OD. For the larger variations in absorbance over the half-power bandwidth, linearity error for on-peak wavelength settings are roughly 1.5- to 2-fold higher than for slope settings, and error for a triangular bandpass is roughly double that for a rectangular bandpass.

These results belie the common notion that densitometric profiles must be scanned at a maximum-absorbance wavelength. Assuming the same range of absorbance values within the bandpass, quantitation at a peak is subject to the effects of reduced absorp-

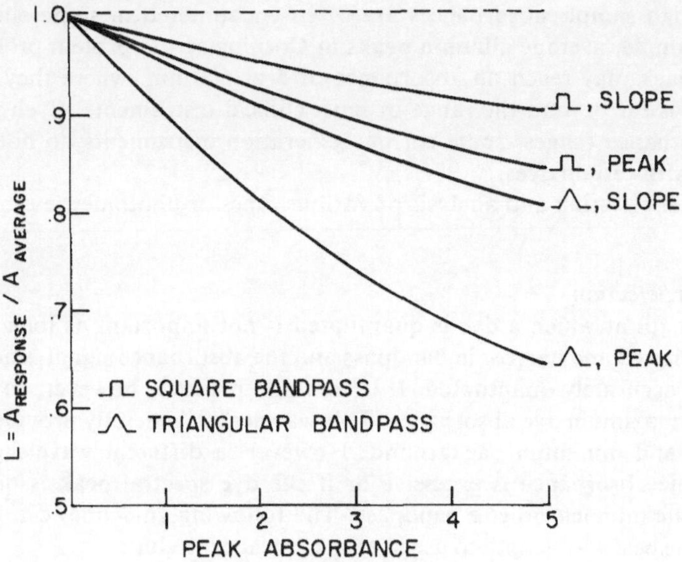

FIGURE 4. Effect of bandpass shape and spectrum shape on linearity factor, $L = A_{RESP}/A_{AVG}$ (from Table 6, $d = 0.5$)

tion on both sides of the central wavelength, while quantitation on a slope is subject to reduced absorption on only one side. In practice, however, this effect is usually more than offset by the generally lesser variation in absorbance at the top of the peak. On the other hand, if accurate as well as linear absorbance values are desired, scanning on a linear slope has the additional advantage that the average absorbance equals the central absorbance. Scanning at an absorbance peak will result in reduced average absorbance in the slit, even if linearity error does not occur (the AR values of Table 5 apply). Scanning on an absorbance slope, however, requires higher wavelength accuracy, as variations in absorbance as large as several percent per nanometer are not uncommon.[20] Again, the particular situation will dictate the best approach.

The following crude guideline for allowed spectral variation can be established, based on the assumption that the triangular and square bandpasses will bracket the more realistic bandpass shapes. If less than about 1% linearity error per OD is desired, sample absorptivity should not change by more than 20% over the half-amplitude densitometric bandpass; if the bandpass is quite square, allowed variation approaches 30%.

Densitometric errors due to inadequate spectral filtering are especially insidious because they may occur without being readily apparent to the operator. Densitometer response to dyes may fall off gradually without reaching full scale on the instrument, so there may be no indication of overrange. Linearity tests with neutral density samples (e.g., photographic step tablets) are inadequate to disclose spectral problems because the filters absorb light at both the primary and interfering wavelengths. The sample may not appear very dense visually because the visual transparency of the sample is due to those wavelengths where the dye does not absorb and is therefore not quantitated. Pinholes in interference filter coatings can significantly degrade performance, particularly if the filter is masked by a small aperture, which enhances the effect. A monochrometer is no guarantee of spectral performance. While a narrow bandpass can readily be achieved with a monochrometer, spurious reflections and stray transmission at harmonically related wavelengths can present a problem.

The above effects are a practical, though little appreciated, problem in the clinical laboratory. High sample absorbances are often encountered in some clinical procedures. For example, average albumin peaks in Corning serum protein profiles stained with amido black may reach an absorbance of 5 at 600 nm, where they are usually scanned. This is far beyond the range of most clinical instruments. Even within their specified absorbance ranges, some current generation instruments do not give linear performance with certain dyes.

A further classification and analysis of various types of photometric error is in Reference 6.

5. Wavelength Selection

The wavelength at which a dye is quantitated is not important as long as the light beam is well filtered and narrow in bandpass and the absorbance signal is not too small or large to be accurately quantitated. It is common practice, however, to scan at the wavelength of maximum dye absorbance. This wavelength generally provides the greatest sensitivity and minimum background. However, a different wavelength may be better if sample absorbance is excessive or if the dye spectral peak is quite narrow compared to the monochrometer bandpass. The following guidelines can be followed for selecting the best wavelength to use in each clinical procedure.

1. Select the wavelength setting which gives the highest absorbance without routinely overranging the instrument. If a spectrum of the dye is not available, relative absorbance as a function of wavelength can be determined by scanning the electrophoretic profile together with a strip of neutral density film (e.g., exposed photographic film). Dye absorbance at each wavelength is indicated by the relative heights of the dye and neutral density peaks.
2. Avoid settings for which dye absorptivity varies more than 20% over the half-amplitude bandpass of the monochrometer filter.
3. Unless instrument has been verified linear for the dye/wavelength combination chosen, perform a dye solution linearity test as outlined in the specifications and testing section.

V. FLUOROMETRIC THEORY

When light energy is absorbed by matter, it can be stored and later re-emitted at a different wavelength. This phenomenon is known as fluorescence or phosphorescence, depending on whether the light re-emission occurs almost immediately or is delayed. The mechanism of fluorescence is as follows. Light consists of a stream of particles or "bundles" of electromagnetic energy. These particles are known as quanta, and the energy of each quantum is uniquely determined by the wavelength of the light. Quantum energy is inversely proportional to wavelength, so that shorter wavelength quanta (e.g., UV light) have higher energy than longer wavelength quanta (e.g., visible light). When a quantum is absorbed by a molecule, the molecule is changed from its lowest energy configuration, known as the ground state, to a new electronic configuration of higher energy, known as an excited state. Some or all of this excess energy may be dissipated by internal conversion to heat or by collisions with other molecules. If, however, the excess energy is not dissipated quickly enough by these mechanisms, the molecule may return to the ground state by emitting a photon with a wavelength appropriate to carry away the remaining extra energy. If this excess energy is less than the energy originally absorbed, as is usually the case, the re-emitted photon will have a longer wavelength. Whether an excited molecule re-emits a photon after absorbing one is a matter of chance, and the probability of re-emission is known as the "quantum efficiency." Quantum efficiency depends upon the type of molecule and its chemical and physical environment.

A fluorometer determines the concentration of fluorescent material in a sample by measuring the amount of light which is re-emitted by the sample when it is allowed to absorb light from an excitation source. Since the number of re-emitted photons is the product of a fixed quantum efficiency times the number of absorbed photons, the relationship between light absorption and sample concentration is what determines how fluorescence changes with concentration. The governing equation of fluorescence is therefore

$$F = \phi I_o (1 - 10^{-ac\ell})$$

where F is the rate at which photons are re-emitted by the sample, ϕ is the quantum efficiency of re-emission, I_o is the rate at which excitation photons enter the sample, a is the absorptivity per unit amount of the fluorescent species at the excitation wavelength, c is the (volume) concentration of the fluorescent species, and ℓ is the path length of excitation light through the sample. This equation can be understood as follows. The term $10^{-ac\ell}$ is the familiar Beer-Lambert expression for light transmittance and gives the fraction of incident light I_o that is not absorbed by the sample. Hence, the rate at which light is absorbed by the sample is $I_o (1 - 10^{-ac\ell})$. This term, multiplied by the quantum efficiency, gives the rate of photon re-emission. The product $A = ac\ell$ is the sample's absorbance at the excitation wavelength and is the parameter quantitated in densitometers.

Much clinical fluorometry is performed on thin transparent films viewed at angles nearly perpendicular to the film surface. For this configuration, it is convenient to think of the sample concentration in terms of the product of c and ℓ which gives the amount of sample per unit area of the film.

The most important implication of the above fluorescence equation is that fluorescence is not, in general, a linear function of concentration, since the fraction of light absorbed by the sample is not linear with concentration. The above equation can be rewritten in a way which shows that fluorescence is, however, linear with concentration when absorbance is low. Taylor series expansion of the term $10^{-ac\ell}$ allows the equation to be expressed as a product of a linear term times a nonlinear deviation factor:

$$F = \underbrace{2.303\,\phi I_o A}_{\text{linear term}} \times \underbrace{(1 - 10^{-A})/2.303A}_{\substack{\text{deviation factor} \\ (\approx 1 - A \text{ if } A < 0.3)}}$$

$$ 2.303\,\phi I_o A \times (1 - 1.15A + 0.88A^2 - \ldots)$$

When absorbance is sufficiently low, the deviation factor approaches unity, indicating that fluorescence becomes linear with absorbance, which in turn is linear with concentration. Above an absorbance of 0.05, deviations from linearity begin to be appreciable (>5%). Intrinsic fluorometric error can readily be estimated for clinical samples by measuring net absorbance at the excitation wavelength and evaluating the corresponding deviation factor.

Other effects may introduce additional nonlinearity. The fluorescence emission beam may be attenuated by sample reabsorption at the emission wavelength. Also, if the excitation and emission beams are at right angles and are sharply defined, the sample may absorb a concentration-dependent portion of the excitation light before it reaches the plane of the emission measurement. These sources of nonlinearity are in addition to the intrinsic nonlinear effect discussed above.

A fundamentally different type of fluorescence measurement can be made by mixing the fluorescing substance to be quantitated with a second substance which absorbs without fluorescent emission. If the fluor and competitor absorbances at the excitation wavelength are A_1 and A_2, respectively, the fluorescent emission is given by:

$$F = \phi \frac{A_1}{A_1 + A_2} (1 - e^{-(A_1 + A_2)})$$

If total absorbance is sufficiently high, all of the light can be considered absorbed, so fluorescent emission is given by:

$$F = \phi \frac{A_1}{A_1 + A_2}$$

If $A_2 >> A_1$, emission will be essentially linear with species 1. This technique allows a linear measurement to be made at high values of A_1 in exchange for reduced signal. It can be shown, however, that at comparable signal levels the standard fluorescence technique provides superior linearity. Aspects of the above analysis apply to surface fluorometry of nontransparent samples, such as certain cellulose acetate powder films.

A main advantage of fluorometry is its sensitivity. It is useful for quantitating samples with absorbance far too low to be accurately measured in a densitometer.

If sample absorbance is either very low or very high, the choice between fluorometry and densitometry is clear cut. Unfortunately, some clinical procedures are conducted in the problematic range where fluorescence is somewhat nonlinear, but absorbance is still too low for a good signal-to-noise ratio. To complicate the matter, certain clinical procedures (e.g., lactate dehydrogenase isoenzymes) are available both as a fluorometric assay and as a colorimetric assay in which the fluorescent species is chemically converted to a visibly colored end product. A choice between these procedures cannot be made on an *a priori* basis. While densitometric quantitation is intrinsically more linear at high concentration, in practice this advantage may be offset by other factors. Low absorbance can increase error due to background and artifacts. The nonlinearity of the basic isoenzyme chemistries is a limiting factor in both types of procedures. The additional reaction converting the fluorescent species to a colored product is a possible source of further nonlinearity and nonspecific background. Fluorometric quantitation allows the use of a long slit that can sweep over the entire width of a sample track. Densitometric slits must be small to insure linear quantitation, so that only a partial cross-section of the sample is scanned. In view of the multiplicity of variables, overall system accuracy should be evaluated on the basis of absolute quantitation of serially diluted samples over the range of clinical interest.

A. Fluorescence Activity

To be useful, a physical measurement should be expressed in units having a fixed and definite meaning. Absorbance and fluorescence measurements are quite different in this regard. For example, the densitometric result, "$A_{260} = 2$", has a fixed, quantitative meaning, and is readily communicated. On the other hand, fluorescence measurements are generally relative in nature and less easily communicated in quantitative terms. A major reason for this difference is that the uncontrolled variables that affect the measurement of I and I_o in densitometry are cancelled when the transmittance $T = I/I_o$ is computed. Variations in source emission and detector sensitivity with wavelength are automatically compensated, since I and I_o are measured at the same wavelength and quite often over the same optical path. Finally, absorbance is proportional to concentration, allowing definition of useful concentration-independent absorbance parameters such as "molar absorptivity". The situation is quite different in fluorometry. Fluorometric output is directly proportional to both excitation intensity and detector sensitivity. These parameters are usually uncontrolled, often unknown, and sometimes unstable; moreover, they are functions of wavelength. Since the excitation and emission wavelengths are different, the source/detector dependence of the fluorescence measurement cannot be readily cancelled. Finally, fluorescence is not linear with concentration over an indefinite range, discouraging the use of concentration-independent parameters.

While the clinical chemist rarely is interested in taking absolute fluorescence spectra, he quite often is interested in making fixed-wavelength fluorescence measurements which he can compare from day to day and from laboratory to laboratory. Such measurement stability is essential, e.g., for routine fluorescent quantitation of "total enzyme" or for development of more sensitive fluorometric isoenzyme procedures. Here a large, uncompensated change in, say, excitation lamp output could be disastrous. For this reason, it is common practice to use relatively stable fluorescent substances such as quinine sulfate as reference standards. This approach, as now practiced, has several limitations. It falls short of assigning numerical values to the measurements themselves. The concentration of the standard, and the relative instrument signal for the sample must both be reported. Secondly, due to possible nonlinear emission of the standard, results obtained relative to standards at different concentrations may not be cross comparable. However, use of a standard above its linear range might be required, e.g., to make its signal comparable to the data being taken. Finally, no convention exists for expressing the linear ranges of fluorometric instrumentation which may exceed the linear emission range of the species that will be routinely measured.

These limitations can be overcome by the following "fluorescence activity" convention.

Fluorescence activity specifies the level of fluorescent re-emission in terms of numerical units derived from the re-emission of a reference fluor at low absorbance levels where fluorescence and concentration are linearly related. The activity units are defined as follows. The fluorescence equation can be rearranged to the form

$$F/I_o = \phi(1 - 10^{-ac\ell})$$

The term on the left gives the level of fluorescent re-emission, representing what might be considered the fluorescence "strength" of the sample. This term is, unfortunately, very difficult to measure experimentally. However, for fixed excitation and emission wavelengths and a given physical state of the sample, the only independent variable on the right side of the equation is the area concentration $c\ell$, and this is easily measured. Hence F/I_o can be specified by an equivalent value of $c\ell$. To make the specification linear and applicable to any level of fluorescence strength, F/I_o is associated with a value of $c\ell$ through extrapolation of the linear-range equation $F/I_o = 2.303\,\phi a(c\ell)$. The value of $c\ell$ associated in this way will be termed the fluorescence activity and be abbreviated FA.

To determine the fluorescence activity of a sample, it is unnecessary to directly measure any of its absorption or emission parameters, since the ratio F/I_o, though unknown, changes linearly with fluorescence activity. If the sample and a fluorescence standard in its linear range are quantitated together in a linear fluorometer, then the instrumental signal for the sample is given by:

$$f_{sample} = k\,(F_{sample}/I_o) = 2.303\,k\,\phi\,a\,(FA)_{sample}$$

and the signal for the standard is given by:

$$f_{standard} = k\,(F_{standard}/I_o) = 2.303\,k\,\phi\,a\,(c\ell)_{standard}$$

where k is some fluorometer-dependent proportionality constant. Hence

$$FA_{unknown} = \frac{f_{unknown}}{f_{standard}}\,(c\ell)_{standard}$$

Note that the fluorescence activity of the standard fluor is equal to its true concentration at low concentrations, but falls below true concentration when fluorescence deviates below linear. Fluorescence activity is therefore quite analogous to the concept of "chemical activity", in which the chemical behavior of a nonideal solute at high concentration is specified in concentration units that are considered equivalent to the chemical behavior of the solute at infinite dilution.

Fluorescence activity is conveniently reported in the format FA = concentration standard (state), e.g., FA = 10,000 ng/cm² quinine sulfate (0.1 $N\,H_2SO_4$). Experimentally, this may correspond to an output signal which is 200 times that of a primary standard consisting of a 1-mm pathlength cuvette containing 500 ng/cm³ quinine sulfate in 0.1 $N\,H_2SO_4$. The fluorescent activity figure serves to compress the related experimental details into a single number.

Because of the linear nature of the activity convention, secondary standards are readily calibrated by scanning them together with the primary standard. The activity of the secondary standard is simply the activity of the primary standard multiplied by the ratio of the instrument outputs for the two standards. We have used fluorescent Corning 0 — 51 glass filters[5] calibrated in NADH units as secondary standards in some of our isoenzyme assay studies.

The fluorescence activity convention is quite useful for establishing a system of fluorescence level specification for fixed wavelength fluorometers of a given type. This system allows quantitative comparisons between results obtained in different laboratories using the same instrument. Since the fluorescence activities of different fluors do not change in fixed proportion when excitation or emission wavelength is changed, the activity convention is less useful for comparisons on variable-wavelength fluorometers or between fixed-wavelength fluorometers of different types. Changes in wavelength or fluorometer type can, however, be accommodated by multiplying the activity values by fixed conversion factors. Fluoroescence activity is useful for specifying instrument ranges, since it can be referenced to substances that the user is likely to quantitate, without being limited to the maximal fluorescence of those substances.

VI. FLUOROMETRIC INSTRUMENTATION AND ERRORS

Though fluorometry is in principle simpler than densitometry, fluorometric performance has often been inferior to densitometric performance in clinical scanners. The following factors are key to good fluorometric performance.

A stable excitation source is essential. Since fluorescent emission is always proportional to excitation intensity, a given percentage of source fluctuation causes the same percentage error. This is in contrast to densitometry, where source fluctuation is significant only at low absorbance levels. The UV-phosphor fluorescent lamps commonly used in clinical fluorometers sometimes flicker or drift in output and should be suspected if reproducibility problems are observed. A gradual, significant loss in lamp output with time can be expected.

Positive sample positioning relative to the excitation source and detector is essential. Excitation intensity generally falls in at least inverse proportion with source-to-sample distance. A few percent difference in this distance usually causes a comparable or larger change in signal. The fluorescent sample should therefore be confined to a plane if possible and not allowed to droop or curl.

Sample temperature and hydration should be uniform. Sample temperature and physical state effect the quantum efficiency by influencing the rates of the energy-transfer processes competing with fluorescent re-emission. The temperature coefficient of fluoroescence varies from nearly 0 to 3%/°C, depending on the species and solvent.[14,15] Notably, the quantum efficiency of NADH increases several-fold when solutions are dried.

In current instrumentation, fluorometric spatial resolution is often poor, due in part to the common practice of locating both excitation and emission optics on the same side of the sample, which necessitates a compromise geometry.

Geometries which allow the sample to reflect excitation light into the detector should be avoided. Due to imperfect filtering, the excitation beam usually contains wavelengths to which the detector is sensitive. If reflection from the sample occurs, waviness or uneven reflectivity of the sample surface may cause uneven background.

Fluorometric slits can and should be long enough to sweep the entire width of fractions. There is no linearity error associated with an excessively long slit. Response along the slit should, however, be uniform.

Fluorometric electrophoretic assays commonly give broader bands than the corresponding colorimetric assays. The fluorescent species in most procedures is NADH which is soluble and diffuses. In colorimetric procedures, the absorbing dye is often insoluble (e.g., formazan) or is bound to a macromolecule and consequently does not diffuse.

VII. INSTRUMENTATION AND PERFORMANCE SPECIFICATION

A. System Description

The following functional description of the Corning 720® fluorometer/densitometer is provided to give a framework for discussing instrument subsystems and performance evaluation. While specific design details may differ, the Corning 720® is similar to other clinical scanners in its basic principles of operation.

The Corning 720® scanning fluorometer/densitometer is designed for the quantitation of clinical electrophoretic separations on nonopaque thin-film media. The electrophoresis sample is placed in a carrier which indexes the various sample tracks under the optics. The operator slides a mode switch to select fluorescence or one of three densitometric wavelengths. The operator inserts a 3 × 8 in. data output card into the instrument and initiates a scan by pressing a paddle-shaped scan switch in a direction marked "auto". A stage bearing the sample and pen recorder servo-mechanism moves to the right, sweeping the sample past the optics. During this forward scan, electric circuits detect and hold the minimum and maximum signal levels from the sample, and the pen recorder draws a baseline on the output card. At the end of the forward scan, electronic gain and baseline of the instrument are automatically set to give maximum and zero deflections of the pen recorder for the stored signal extremes. The stage reverses, and the recorder traces the analog signal profile on the output card. An automatic valley detector selects integration limits by detecting changes in the signal that correspond to the boundaries of the various fractions. These integration limits are marked by small blips on the output profile. At the end of the scan, a light-emitting-diode (LED) display shows the number of sensed fractions. The operator depresses a button which cycles the display to show the percentage of total area for each fraction.

If the operator prefers to select different integration limits, vertical pencil lines are marked under the desired positions on the output card, and the scan switch is pressed in the opposite direction, marked "manual". The pencil marks are detected during the ensuing scan by an IR "eye", and are used as integration limits. Area data appears, as before, in the LED display. A microprocessor controls all instrument mechanical and electronic sequencing and processes the data. The microprocessor also detects various operator and instrument fault conditions and if necessary sends diagnostic error codes to the display. A separate, microprocessor-controlled printer module is available. During or after the scan, the operator enters patient identification, assay total value, and number-coded comments, if any. The output card with the analog trace is inserted in the printer, and the printer outputs time, date, operator-entered data, test type, fraction data in percentage and absolute terms, and the results of operator-programmed peak ratio computations. The unit is interfaceable to a central hospital computer.

One point of functional difference between the Corning 720® and other scanners is auto-zero. Most other scanners set baseline to the signal level at the beginning of the scan, while the Corning 720® sets baseline to the minimum signal level encountered.

A block diagram of portions of the Corning 720® optical and mechanical system is in Figure 5. These function as follows. In fluorometric mode, UV light from a pair of UV-phosphor fluorescent lamps passes obliquely through a UV-passband filter and illuminates the sample. Fluorescent emission from the opposite side of the sample passes through an objective lens and three blocking filters. The objective lens serves to image the sample plane onto a secondary aperture which defines the fluorometric slit. Emission light passing through the secondary aperture is collected by an aspheric condensing lens which images the exit pupil of the objective lens onto a silicon PIN photodiode. The photodiode converts the incident light into an electrical signal. In densitometric mode, light from an incandescent lamp passes through a lens and bandpass filter and becomes imaged upon a narrow, slit aperture. A film transport stage holds the sample in very close proximity to the slit and scans the sample past the slit at a uniform rate. Light transmitted by the sample is collected by the objective lens and subsequent optics as described for fluorometric mode. In densitometric mode, the secondary aperture, though larger than the image of the illuminated area of the sample, serves to block residual low-level light far from the image center.

A block diagram of relevant electronic subsystems of the Corning 720® is in Figure 6. Light detection and initial signal processing are performed by an integrated preamplifier stage. Light striking the photodetector is converted by the detector and a primary amplifier to a voltage which is proportional to light intensity. In fluorometric mode, the output of the primary amplifier is amplified further by a second linear amplifier. This amplifier is connected to the preamplifier stage output in fluorometric mode and provides a signal which changes linearly with sample fluoresence. In densitometric mode, the output of the primary amplifier is processed by a logarithmic amplifier which changes in output by a fixed increment for each decade-factor change in the input voltage. The logarithmic amplifier is connected to the preamplifier output in densitometric mode and provides a signal which changes linearly with sample absorbance. Hence, in either fluorometric or densitometric mode, the preamplifier output is a voltage which changes linearly with the optical properties of the sample. This output voltage, however, includes a baseline offset which depends on variable electronic and optical parameters and upon background fluorescence or absorbance of the sample.

The preamplifier output is subjected to further processing which eventually results in data output from a pen recorder and digital integrator. The preamplifier output

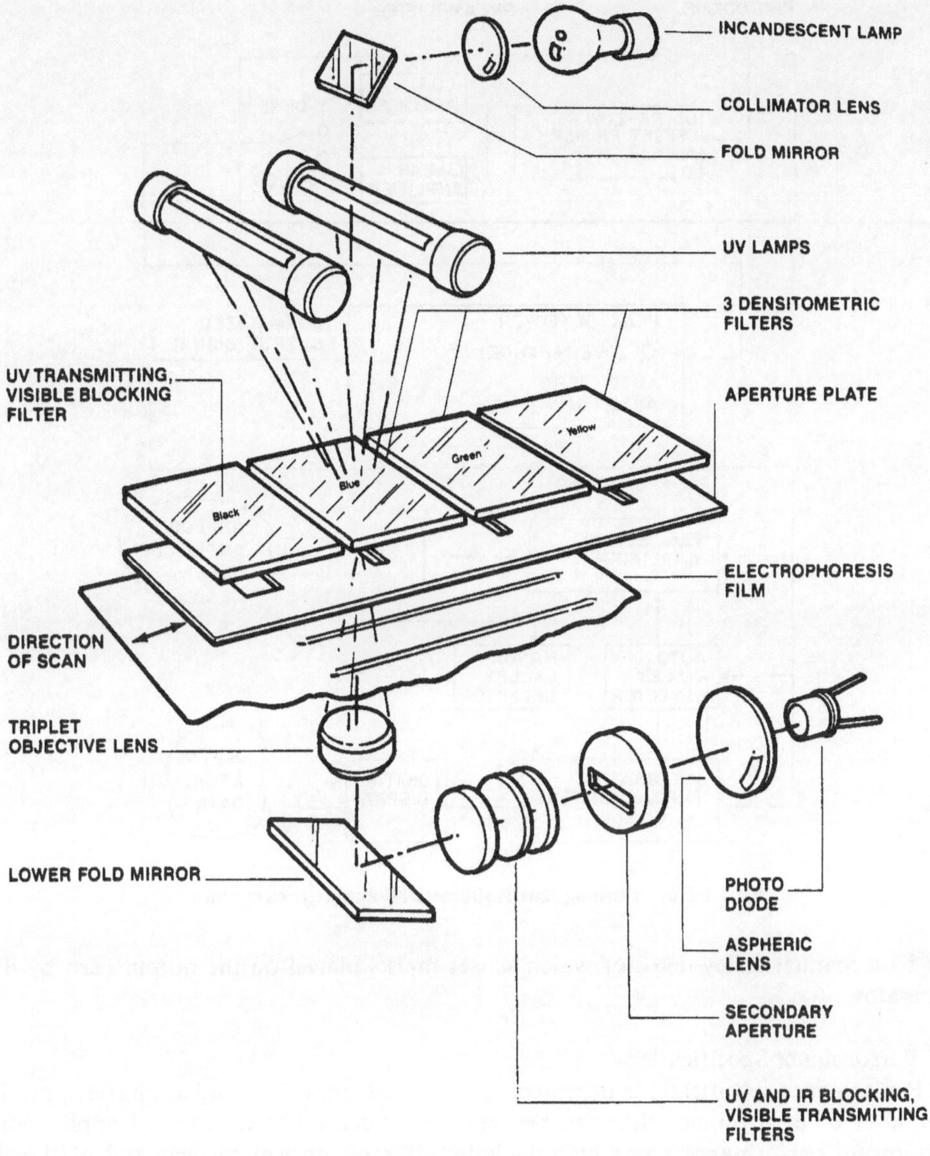

FIGURE 5. Corning 720 fluorometer/densitometer optical system.

first passes through auto-zero and auto-gain stages which offset and rescale the input signal, so that the minimum input signal during the scan gives a zero output, and the maximum input signal gives a "full scale" output to the subsequent pen servo and integrator stages. This off-setting and rescaling process is termed "normalization" of the signal. The normalized signal (auto-gain output) is fed to circuitry driving the pen recorder which traces a profile corresponding to net density or fluorescence along the sample and to a digital integrator which determines the fractional area under the profile for selected portions of the scan. The integrator consists of a voltage-to-frequency converter (VFC) which generates pulses at a rate proportional to the input signal and the microprocessor which counts pulses occurring between selected integration limits and computes the fractional area data for display. Integration limits are provided to the microprocessor by the auto-valley detector which senses slope changes in the signal

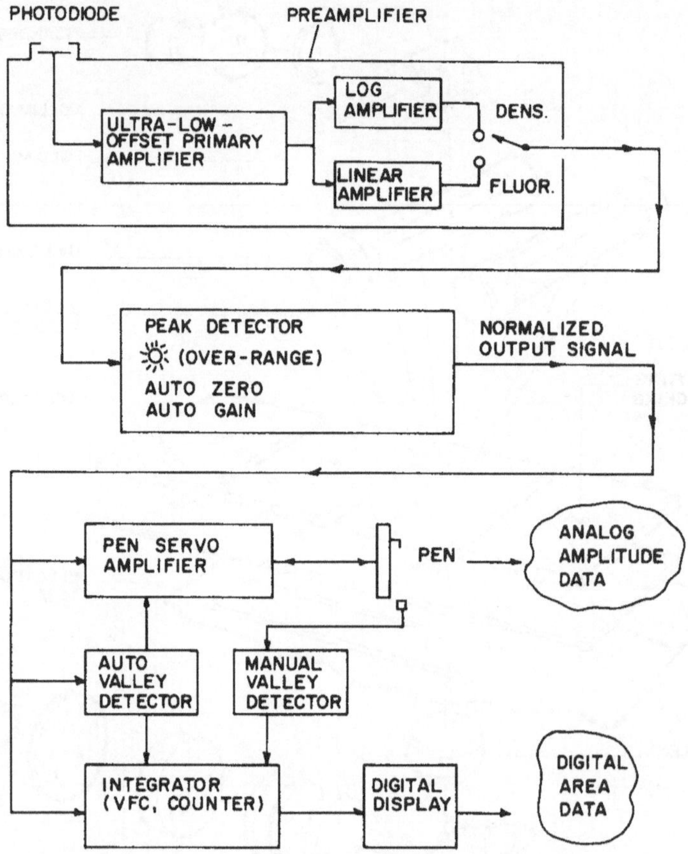

FIGURE 6. Corning 720 densitometer/fluorometer electronics.

or by a manual valley detector which senses marks placed on the output card by the operator.

B. Performance Specifications

Performance specifications provided for a clinical scanner should accurately portray the level of performance that can be expected under a broad range of application. Important performance parameters include: effective optical slit size and off-image blocking, optical/electronic rise time, optical and electronic linearity, linear range, auto-zero and auto-gain accuracies, pen recorder accuracy, electronic integrator accuracy, and scan rate uniformity.

Definition and measurement of these parameters will now be described.

1. Range

An instrument range specification should show the minimum and maximum inputs which can be accurately quantitated. If an instrument has fixed gain or range settings, these can be enumerated. If the instrument has variable or automatic gain and baseline, the maximum and minimum settings for range and baseline should be indicated. An alternate means of doing so is to specify the maximum and minimum signals which can be accurately quantitated. For example, the lower range limit or sensitivity can be defined as the input increment which gives a 1% of full-scale response at maximum instrument gain, and the upper range limit can be defined as the maximum input level for which a linear output response is obtained or confirmed. In the latter type of specification, it is assumed that the instrument baseline can be set anywhere between zero and the upper range limit.

Densitometric range can be indicated in absorbance units. Fluorometric range can be indicated in fluorescence activity units of a common reference compound, preferably one which is clinically significant (e.g., NADH) or in common use (e.g., quinine sulfate). A listing of ranges for the various most commonly quantitated species would be even more informative.

2. Linearity

A specification of a linear range is absolutely meaningless unless some criterion of accuracy is specified. If the densitometer reads directly in absorbance units or has fixed gain ranges, the maximum absolute deviation in output from the correct absorbance value can be specified. If the instrument has variable or automatic gain or if the output is not calibrated in absolute absorbance units, a meaningful linearity specification is harder to define. For example, the Corning 720® is intended for determination of relative profile amplitudes and area over a very wide range of input conditions. If a linearity specification for this type of instrument is to be applicable to output data from an individual scan, it must refer directly to the limited input range encountered during the scan. For this reason, performance accuracy should be specified as a given percentage of the output range for any input profile lying within the specified range limits.

Directly verifying this type of specification over a representative range of inputs is a formidable problem. One approach is to specify and verify input/output linearity in two more easily tested stages. An internal instrument signal representing density or fluorescence is used as an intermediate parameter. For example, in the case of the Corning 720®, the preamplifier output voltage is a convenient intermediate parameter. The first-stage specification, optical and electronic linearity, then refers to the linearity with which increments in sample density or fluorescence are represented by increments in preamplifier output voltage. The second-stage specifications, pen recorder accuracy and electronic integrator accuracy, refer to the accuracy with which voltage increments at the preamplifier output are converted to proportional pen displacements or area-acquisition rates, respectively. These specifications are more easily verified over a representative variety of input ranges. As will be described later, total input/output error is less than the sum of the errors for the two specification stages.

Optical/electronic linearity can be measured by recording preamplifier output response to a series of fluorescent or optically dense samples. A regression line is fitted to the input/output data by adjusting the line's slope and intercept to minimize the deviations from the line. The easiest way of adjusting the line is via least squares linear regression, which minimizes the sum of the squared deviations from the line. A second, more complicated method, is to adjust the line to minimize the largest deviation from the line. The latter method gives a somewhat better linearity figure. Adjusting the line intercept is appropriate if the intermediate signal is followed by an automatic or manual zeroing stage which removes baseline offsets due to sample or instrument background. Linearity is specified in terms of the maximum deviation of data points from the regression line, expressed as a percentage of the difference between the largest and smallest datum (full scale).

3. Pen Recorder Accuracy

Pen recorder accuracy refers to the accuracy with which net changes in the intermediate density or fluorescence signal are represented by proportional excursions of the recorder pen. Adequate pen recorder accuracy is required to insure that the analog output is a faithful reproduction of the net absorbance or fluorescence profile of the sample and to insure that accurate determinations of relative peak amplitude can be made from the output trace. Recorder accuracy can be measured by introducing a range of test signals at the preamplifier output level. Gain and zero for this input range are set by the usual manual or automatic procedure with the minimum input signal being used to set baseline. Pen displacements corresponding to the input levels are measured from the chart baseline. Data for pen displacement vs. injected signal is fitted by a zero-intercept regression line which specifies that the pen should be at baseline for the minimum input signal. A zero-intercept regression is appropriate, since this test includes the instrumental zeroing stage. For accurate representation of the signal, the baseline input signal must be referenced to the recorder baseline. Deviations of the pen recorder positions from the regression line are expressed as a percentage of the full-scale recorder displacement.

4. Integration Accuracy

The analog output of a scanner generates a curve representing the relative distribution of detected fluorescence or absorbance along the scanned electrophoretic sample. To determine the relative amounts of material in discrete bands or peaks of the profile, it is necessary to spatially integrate the profile to determine the relative area under different sections of the curve.

In most instruments, spatial integration is accomplished by feeding the normalized analog output signal to an electronic integrator which integrates the signal with respect to time. The integrator usually consists of a voltage-to-frequency converter which generates pulses at a rate proportional to the signal and a counter which counts the number of pulses that occur while different regions of the profile are scanned. Manual "zig-zag" integrators work the same way, except that the operator does the pulse counting. To the extent that the sample is scanned at a uniform rate, the integrator gives an output equivalent to integration with respect to distance along the profile. Two factors therefore determine the accuracy of spatial integration: electronic integrator accuracy (proportionality between integrator pulse rate and net signal) and scan rate uniformity.

5. Electronic Integrator Accuracy

Electronic integrator accuracy can be determined by using injected signals to determine the proportionality with which voltage increments representing net fluorescence or absorbance at the intermediate signal stage are converted to an area-acquisition (pulse) rate at the counter. Gain and zero for the desired input range are set by the usual instrument procedure. Pulse frequency is then measured as a function of voltage increment about baseline. A zero-intercept regression line is fitted to the data, and the largest deviation in pulse rate from the regression line is determined as a percent of the maximum pulse rate. A zero-intercept line must be fitted, since for accurate integration, the integrator output must be proportional, not just linear, with respect to the net input signal.

Scan-rate uniformity refers to the uniformity with which the sample is swept past the optical slit. Scan rate uniformity can be expressed as the greatest deviation in scan rate from the median scan rate, divided by the median scan rate. Nonuniform scan rate may or may not affect the analog output. If the recorder pen is directly coupled to the stage which carries the sample, as on the Corning 720®, a nonuniform scan rate will not affect the recorder output profile, even though integration results are affected. If a separate chart drive is employed, both scan and chart drive nonuniformities will affect the results. Usually this is a quite minor concern, since area information is generally obtained from the integrator, not the recorder output.

Optical/electronic errors and pen recorder/integrator errors can be combined by addition to give an overall linearity figure. This follows from the fact that output data can be correlated with the input data by superposition of the regression lines used to fit the two sets of data to the intermediate signal parameter. Maximum deviation of input/output data from the superimposed line is no greater than the sum of the individual maximum deviations. True I/O deviation is likely to be even less than this sum, since the maximum deviation for each regression may not coincide and because the superimposed line is probably not the best fit to the combined I/O data.

It should be noted that the correlation coefficient between input and output data is an extremely insensitive indication of linearity error. For example, if sufficient stray light exists to give a −10% deviation from linearity at 4 OD, a zero-intercept regression line will show a deviation of −5.2%. However, the correlation between measured absorbance and concentration is still 0.998, an impressive number which gives the illusion of extremely high linearity. The correlation coefficient is poorly suited for the analysis of consistent errors such as deviation from linearity, but is widely used for this purpose.

6. Auto-Zero and Auto-Gain Accuracies

Zeroing accuracy refers to the accuracy with which the minimum or baseline signal of a profile is instrumentally offset to provide a zero level output from the integrator and pen recorder. Zeroing accuracy is expressible in terms of the full-scale output signal and can be defined as the minimum output signal (residual zeroing error) divided by the maximum output signal (full scale).

A zeroing error changes the amplitude of all peaks by the specified fraction of full scale, and changes the area of all peaks by the specified fraction of the full-scale output area, distributed in proportion to peak width. The effect of zeroing error is most significant in changing the relative area percentages in a profile composed of narrow, high peaks mixed with broad, low peaks (e.g., serum proteins).

Note that the recorder and integrator linearity specifications discussed in the preceding section included the effects of zeroing error. As long as some stage of the I/O linearity analysis involves a zero-intercept regression of baseline-adjusted data, a separate "zeroing-accuracy" specification need not be given.

Auto-gain accuracy, if applicable, represents the consistency with which input profiles are normalized, so that the maximum signals of the profiles all give the same high level output to the electronic integrator and pen recorder stages. Auto-gain accuracy can be expressed as the greatest difference between a full-scale output and the median full-scale output for a representative distribution of input ranges, divided by the median full-scale output.

As long as the operating limits of the pen recorder and electronic integrator are not exceeded, auto-gain errors do not affect quantitation, since all peaks are rescaled to the same degree. However, high auto-gain accuracy facilitates measurement of relative amplitudes from the recorder profile and insures that multiple scans of the same profile will be superimposed.

7. Effective Optical Slit and Off-Image Blocking

The optical slit is the instrumental "window" defining the portion of sample being quantitated. The geometry of the quantitation window can be established in several ways. A physical aperture can be placed very close to the sample. A collimated light beam can be delimited by an aperture some distance from the sample. A lens can be used to focus an image of the sample onto a physical aperture in the optical path. The incident or departing light beam, or both, can be shaped as required to provide resolution. These methodologies are all subject to the effects of diffraction, beam spreading, magnification, and defocusing. Hence, a meaningful resolution figure can only be given by specifying system response at the sample plane. An "effective optical slit"

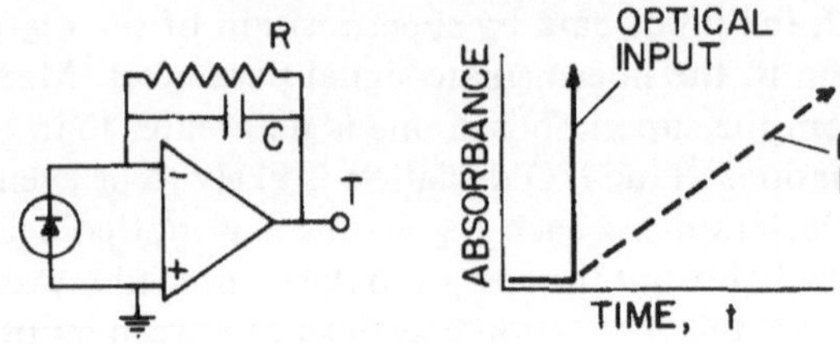

ANALOG SIGNALS FOR:

TRANSMITTANCE $T = T_0 e^{-t/RC}$ (STEP RESPONSE)

ABSORBANCE $A = -\log T$

$= -\log T_0 + t/2.303\ RC$

⟶ RISE TIME OF 2.303 RC PER ABSORBANCE UNIT

FIGURE 7. Response of noise-smoothing RC circuitry to infinite density step.

can be expressed as the length and width of the sample-plane zone giving half-amplitude response. Since blocking of peripheral light is critical in densitometry, slit specifications for off-image blocking should also be given. These can be expressed as the dimensions of the sample area containing all but 10^{-1}, 10^{-2}, 10^{-3}, 10^{-4}, etc. of the detected incident beam power. These dimensions are easily determined by scanning of simple test patterns.

8. Optical/Electronic Rise Time

In order to attain the resolution promised by the dimensions of the optical slit, the optical/electronic response of a scanning instrument must be sufficiently fast to follow the maximum rate of change in absorbance or fluorescence that can occur as a consequence of slit movement over the sample. The time base for sweeping a slit across a boundary between adjacent areas is given by the slit width divided by the scan rate. Rise time can limit the performance of an instrument with high resolution and rapid scan rate.

In densitometric instrumentation, the main determinant of rise time is usually the "RC" time constant of the linear circuitry converting the current from the photodetector into an amplified voltage signal (see Figure 7). Smoothing is introduced at this stage to suppress noise which would otherwise bias the output of the logarithmic amplifier. When an infinite density step is applied (i.e., photodiode current drops abruptly to zero), the linear amplifier output decreases as $V_o e - t/RC$, and the output of the logarithmic amplifier increases as $-\log V_o + t/2.303$ RC. Hence time $t = 2.303(RC)A$ is required for an upward absorbance change of A. The response to a decrease in absorbance is generally much faster. General equations governing transient response are the following. For a linear circuit with time constant RC, the output response for an input step from transmittance T_o to T_1 is given by

$$T = T_1(1 - e^{-t/RC}) + T_o e^{-t/RC}$$

which implies that output T is reached in time

$$t = 2.303\ RC \log [(T_o - T_1/(T - T_1)]$$

The rise time from 10 to 90% of the densitometric range can be calculated using this

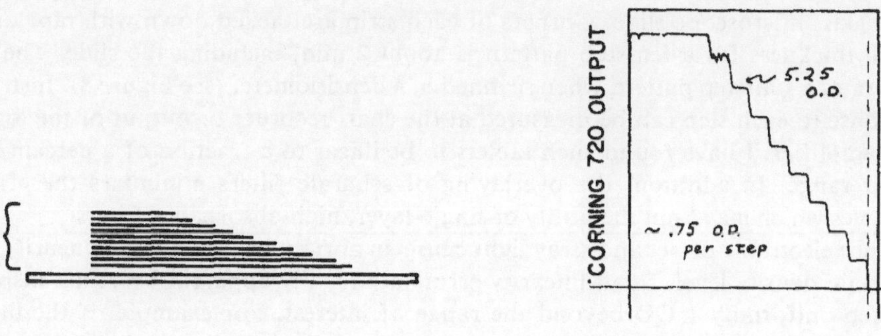

FIGURE 8. Step tablet of overlapped Kodak Wratten Neutral Density Gelatin Filters, Type 96 (0.7 neutral density).

equation. The above equations are also applicable to computation of fluorometric response time. A 10 to 90% rise or fall of a linearly processed signal, based on exponential decay, requires time t = 2.303 RC log (0.9/0.1) = 2.2 RC. Rise times based on circuit analysis should be confirmed experimentally, as secondary effects such as dielectric polarization of smoothing capacitors can greatly increase response time at high absorbance levels.

VIII. PERFORMANCE TESTING AND QUALITY CONTROL

Direct tests of instrument performance are useful to both the instrument designer and clinical user. Clinical quality control procedures based on control sera are not adequate as instrumental tests, since these procedures are generally more variable than the instrumentation, and the results may also simply increase confidence in a consistently erroneous performance. Most of the following tests are simple enough to be performed in the clinical laboratory.

A. Linearity, Linear Range

1. Densitometric Linearity

Two types of linearity should be distinguished: linearity with neutral density samples which absorb equally at all wavelengths and linearity with colored dyes. Tests with calibrated neutral density filters are rapid and easy to perform, but are sensitive only to electronic offset and nonspectral stray light. Neutral density tests will not reveal spectral deficiencies causing nonlinear quantitation of clinical dyes. They are, however, useful for detecting electronic malfunction or misalignment.

Neutral density photographic step tablets are supplied by Kodak and some densitometer manufacturers. A nominal step precision of 5% is common. The National Bureau of Standards Diffuse Density Standard SRM 1008 is a similar product calibrated to a higher accuracy (<1%). These tablets are useful, but have some disadvantages. Certain tablets are too large to be scanned in some instruments. Also, the tablets scatter light strongly and give different results in instruments measuring diffuse and specular density. In particular, the NBS tablet is calibrated in diffuse density. Nonscattering evaporated metal-on-glass (inconel) tablets are available, but are expensive. Also, since these filters attenuate the incident beam by reflecting rather than absorbing it, spurious results due to enhanced stray light can be obtained.[8,12]

A very inexpensive, but very accurate, nonscattering step tablet of convenient dimensions can be made from Kodak Wratten® type no. 96 neutral density filters. Strips of different lengths are cut from the single filter and are stacked in stairstep fashion on a glass microscope slide. Corners of each strip are tacked down with tape or glue. Total thickness for a ten-step pattern is about 2 mm, including the slide. The tablet generates a stairstep pattern when scanned in a densitiometer (see Figure 8). Instrument response to each step can be measured at the chart recorder or output of the logarithmic amplifier. I have found such tablets to be linear to a fraction of a percent over a 5-OD range. In addition, the overlaying of separate filters minimizes the effect of pinholes which may limit the utility of single-layer, high-absorbance filters.

Since electronic offset and stray light cause an abrupt deviation from linearity above a certain density level, small linearity errors are readily apparent by visual inspection of step uniformity 1 OD beyond the range of interest. For example, if the linearity error due to offset is ±0.3 OD at the 4-OD level, the linearity error will be only +0.04, −0.02 OD at 3 OD.

It has been pointed out that satisfaction of Beer's law is not a proof on linear photometric response.[7,17] For example, if photometric response follows the power law, $T_{measured} = (I/I_o)^\gamma$, then the log amplifier response will still be linear, but with altered slope, i.e., $A_{measured} = -\gamma \log (I/I_o)$. This type of effect, however unlikely, can be excluded by making sure that the quantitation of neutral density samples is not only linear, but also gives the correct density values.

Since spectral problems are often the most significant source of nonlinearity, linearity should also be tested using the dyes that are routinely quantitated (amido black 10B, ponceau S, fat red, etc.). Acetylacetone is a convenient, low-volatility solvent for fat red, which has a sharp absorbance peak near 520 nm. The dyes xylene cyanole FF and methyl orange, though not used clinically, have sharp absorbance bands and provide stringent tests of linearity near 600 and 450 nm, respectively. A very accurate dye dilution series can be rapidly prepared by dispensing concentrated dye and blank solutions into vials and weighing the amounts dispensed. A thin (1-mm pathlength) spectrophotometer cuvette is filled with each solution and is scanned in the densitometer. The cuvette should have a bandpass extending into the UV and IR ranges (e.g., quartz or Spectrosil® cuvette). Instrument response vs. dye concentration can be plotted and fitted by linear regression to determine deviations. Since some sources of densitometric error have the greatest relative effect at very low absorbance, a low-range linearity test should also be performed. Amido black dye dilutions scanned at 450 nm can be used to verify linearity to better than 1% over the 0- to 0.1-OD range. Only a single dye/wavelength combination need be tested, since spectral errors have negligible effect in this absorbance range. A major source of error will be light source stability which can be independently checked by observing baseline stability while scanning an empty track at maximum gain.

If only a screening procedure is desired, the following simpler test can be used to detect dye quantitation nonlinearity. The test should be performed for all dye/wavelength combinations that are used.

1. By scanning in a thin spectrophotometer cuvette, determine dye absorbance at a low concentration where the instrument can be presumed linear.
2. Prepare a solution calculated to have the maximum absorbance specified for the instrument (e.g., A = 4).
3. Dilute the above solution by a factor of precisely two.
4. Measure the relative absorbances of the two solutions in the densitometer (A_1, A_2). The relative absorbances can be determined, if necessary, from the graphical output.
5. Compute the fractional deviation from linearity $= (A_1 - 2A_2)/2A_2$. For reasonably accurate quantitation, this should not exceed 0.04.

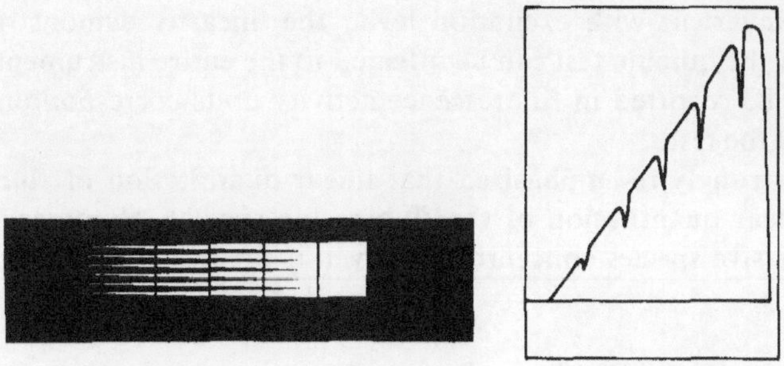

FIGURE 9. Fluorescence linearity test.

Another linearity test is even simpler, though less informative. Linearity errors generally diminish as overall profile absorbance is reduced. Hence, linearity can be checked by seeing if the fraction values are changed when the instrument is switched to a wavelength where absorbance is less, but still significant. If the instrument gives markedly different fraction values at different wavelengths, its performance at one of these wavelengths is not linear. With amido black, results should be similar at 600, 520, and 450 nm. Ponceau S results should be similar at 520 and 450 nm.

2. Fluorometric Linearity

Fluorometric linearity refers to the relationship between light emitted by the sample and the resulting response of the instrument. Direct verification of fluorometric linearity using fluorescent chemicals in solution is complicated by the fact that sample emission is not linear with concentration at high signal levels. For example, quinine sulfate emission may depart significantly (>1%) from linearity above 100 ng/cm^2 concentration. Above this concentration, quinine as well as instrument nonlinearity is being measured. The following two approaches have been used successfully to demonstrate linearity.

Assuming a fixed quantum efficiency, emitted light is proportional to absorbed light. The fraction of incident light absorbed by a given concentration of sample can be computed from the instrument emission spectrum and sample absorption spectrum. A regression line can be fitted to data for instrument response vs. the calculated fraction of light absorbed by the sample. Assuming proportionality between absorbed and emitted light, fluorometric error is indicated by deviation from the regression line. We have used this approach to demonstrate linear quantitation of light emitted by quinine sulfate up to concentrations of 100,000 ng/cm^2. This approach is somewhat awkward, however, since it is indirect and presumes a knowledge of the spectral properties of the fluorometer.

Fluorometric linearity can be more directly verified by the following combination of two complementary tests. In the first test, linearity is demonstrated over a range of quinine concentrations sufficiently low to be linear in emission. In the second test, a highly fluorescent film is placed under the UV excitation filter, and a geometric test mask is placed in the sample carrier. The mask has a series of parallel interrupted wedge apertures that control the level of fluorescence transmitted from the fluorescent film to the detector. When scanned, the mask generates a ramp with fractions of increasing amplitude and area (see Figure 9). The level of UV excitation is adjusted using neutral density filters to give a series of full-scale fluorescence values that span the entire fluorometric range of the instrument. If quantitative results from the mask test are acceptably invarient with excitation level, the linearity demonstrated in the restricted range of the quinine test can be inferred to the entire instrument range. Instrument range can be reported in fluorescence activity units corresponding to the signal level extremes of the test.

It should be strongly re-emphasized that linear quantitation of fluorescence is no guarantee of linear quantitation of the fluorescing species. Nonlinear results can be obtained at excessive species concentration, even if the instrument is operating within range.

B. Auto-Zero, Auto-Gain, Pen Recorder, and Integrator Accuracies

Auto-zero, auto-gain, pen recorder, and integrator accuracies can be evaluated using test signals injected at the circuit level where fluorometric or densitometric linearity is measured. A representative range of test signals should be used. Optical/electronic rise time can be measured by rapid opening and closing of a shutter inserted in the optical train. These tests are straightforward and will not be discussed further.

C. Scan Rate Uniformity

Scan rate uniformity can be tested by scanning a series of uniformly spaced lines on a photomask and measuring time between optical pulses. Alternately, a constant signal can be injected into the integrator, while the mask pulses are used to select fraction valleys. Scan uniformity is determined from the uniformity of the integrator fractions which result.

D. Slit Response Profile

The response profile along the length of the slit can be visualized by scanning a test pattern consisting of a fine, oblique line (see Figure 10, Feature H). Scanning an oblique line is equivalent to sweeping a fixed signal along the length of the slit, and response as a function of position can be read from the resulting recorder output. An effective half-power slit length can be specified on the basis of this measurement.

Figure 10 also shows a pattern providing a test of densitometric resolution (see Feature G). In the pattern, a series of fine lines are spaced in geometric progression. The geometric progression implies that the spacing increases linearly with distance along the pattern. Under most conditions, slit response as a function of position is indicated by the outer envelope of pen excursions generated by scanning the pattern. Where the lines are spaced by more than the half-power width of the slit, the lower limit of the pen excursions gives the response when the slit is centered midway between the lines, provided that the slit response profile is upward concave at these spacings. Thus, the pen excursion envelope indicates response to a fine line as a function of its distance from the slit center. Pen excursions reach baseline at a line spacing where the centered response falls to zero. A node in pen excursions occurs when the lines are spaced by the half-amplitude width of the slit. If a complete null is reached at the nodal position, the rate of change in slit response is symmetric about the half-amplitude positions. With a sharply focused rectangular slit, the transition from nodal to full baseline excursions will be abrupt. Since the overall absorbance of the fine line pattern is very small, the output response is linearly related to the slit sensitivity profile. Because the width measurement is made at a response node, the result is not affected by electronic or pen recorder time constants.

A similar pattern of fine fluorescent lines is useful for measuring fluorometric resolution. Such patterns can be easily fabricated on photographic transparency film, which conveniently provides its own fluorescence. Manufacture and use of test patterns based on the above principles are restricted by patents.[21]

While the above pattern is sufficient in fluorometric mode, an additional resolution

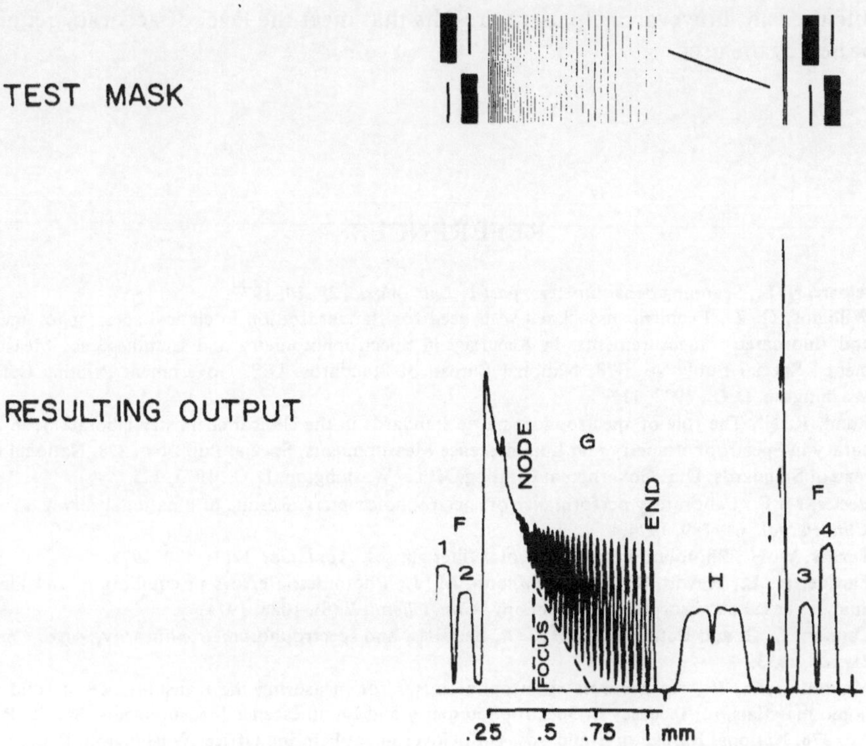

FIGURE 10. Pattern testing resolution, alignment, and response profile of densitometric slit.

characteristic is important in densitometry. Response to high densities is limited by both densitometric rise time and residual, low-level response to light far from the slit center. The pattern discussed above is insensitive to these effects. Due to different degrees of off-image blocking, slits having essentially the same low-density resolution can have radically different resolutions at high density. A Sayce pattern[18] is useful for determining high-absorbance resolution. The Sayce pattern consists of alternating opaque and transparent zones, spaced in geometric progression. When a Sayce pattern is scanned, the response to an opaque block of given width is limited to a certain density level by rise time and stray slit response. Densitometric response as a function of block width can be determined directly from the Sayce pattern output and can be cited as a resolution parameter. For example, "4-OD resolution = 2 mm". For slow scan rates, this would imply that 0.0001 of the slit response occurs further than 1 mm from the slit center. Results can be presented in tabular form or as a plot of resolution versus absorbance.

IX. CONCLUSION

While materials and methodology usually are the major sources of variability in electrophoretic results, the view that instrumental sources of error are always negligible is an unwarranted assumption. Existing clinical instruments differ significantly in important parameters, such as resolution, linearity, and range. Malfunction and incorrect use of instrumentation can also contribute to error. Currently the clinical user is in a poor position to evaluate the quantitative performance of an instrument, either directly or through published information. Specifications, performance testing, and knowledgeable use can, however, help assure results that meet the level of accuracy required or expected by the user.

REFERENCES

1. **Alpert, N. L.,** Scanning densitometers, part 1, *Lab. World,* 28, 20, 1977.
2. **Williams, G. Z.,** Problems associated with need for standardization in clinical spectrophotometric and fluorometric measurements, in Accuracy in Spectrophotometry and Luminescence Measurements, Special Publ. No. 378, National Bureau of Standards, U.S. Government Printing Office, Washington, D.C., 1973, 117.
3. **Rand, R. N.,** The role of spectrophotometric standards in the clinical chemistry laboratory, in Accuracy in Spectrophotometry and Luminescence Measurements, Special Publ. No. 378, National Bureau of Standards, U.S. Government Printing Office, Washington, D.C., 1973, 125.
4. **Beeler, M. F.,** Laboratory performance of spectrophotometers. Results of a national survey, *Am. J. Clin. Pathol.,* 61, 789, 1974.
5. **Turner, W. H.,** Photoluminescence of color filter glasses, *Appl. Opt.* 12(3), 480, 1973.
6. **Pardue, H. L., Hewitt, T. E., and Milano, M. J.,** Photometric errors in equilibrium and kinetic analyses based on absorption spectroscopy, *Clin. Chem.,* 20(8), 1028, 1974.
7. **Cannon, C. G. and Butterworth, I. S. C.,** Beers law and spectrophotometric linearity, *Anal. Chem.,* 25, 168, 1953.
8. **Mavrodineanu, R.,** An accurate spectrophotometer for measuring the transmittance of solid and liquid materials, in Accuracy in Spectrophotometry and Luminescence Measurements, Special Publ. No. 378, National Bureau of Standards, U.S. Government Printing Office, Washington, D.C., 1973, 31.
9. **Hardy, A. C. and Young, F. M.,** The correction of slit width errors, *J. Opt. Soc. Am.,* 39, 265, 1949.
10. **Eberhardt, W. H.,** Slit-width effects in spectrophotometry, *J. Opt. Soc. Am.,* 44, 22, 1954.
11. **Broderson, S.,** Slit width effects, *J. Opt. Soc. Am.,* 44, 22, 1954.
12. **Mavrodineanu, R.,** Considerations for the use of semi-transparent metallic thin films as potential transmittance standards in spectrophotometry, in Standardization in Spectrophotometry and Luminescence Measurements, Special Publ. No. 466, National Bureau of Standards, U.S. Government Printing Office, Washington, D.C., 1977, 127.
13. **Beeler, M. F. and Lancaster, R. G.,** Survey to assess the extent of stray light problems in precision spectrophotometry, *Am. J. Clin. Pathol.,* 63 (Suppl.), 953, 1975.
14. **Elevitch, F. R.,** Operational Variables in Fluorometry, *Fluorometric Techniques in Clinical Chemistry,* Little, Brown, Boston, 1973, 41.
15. **Udenfriend, S.,** Quantum yield and lifetime of fluorescence, *Fluorescence Assay in Biology and Medicine,* Vol. 2, Academic Press, New York, 1969, 19.
16. **Coates, V. J. and Hausdorff, H.,** Interferometric method of measuring the spectral slit width of spectrophotometers, *J. Opt. Soc. Am.,* 45, 425, 1955.
17. **Hawes, R. C.,** Technique for measuring photometric accuracy, *Appl. Opt.* 10(6), 1246, 1971.
18. Teledyne Gurley, Precision Resolution Targets, Bulletin 8000, Troy, N. Y., 1974.
19. **Rand, R. N.,** Practical spectrophotometric standards, *Clin. Chem.,* 15, 839, 1969.
20. **Burke, R. W., Deardorff, and Menis, O.,** Liquid absorbance standards, in Accuracy in Spectrophotometry and Luminescence Measurements, Special Publ. No. 378, National Bureau of Standards, U.S. Government Printing Office, Washington, D.C., 1973, 95.
21. **Brezinski, D. P.,** U.S. Patent 4,097,735, 1978. (foreign patents pending).

AUTOMATED LEUKOCYTE DIFFERENTIAL ANALYZERS

Charles N. Carter

TABLE OF CONTENTS

I. Introduction 36

II. Leukocyte Differential Count 36

III. Historical Background 37

IV. Pattern Recognition vs. Histochemical Techniques 39

V. Automation 39

VI. Hardware Subsystems of an Automated Leukocyte Differential Analyzer 40

VII. Image Processing and Feature Extraction 44

VIII. Pattern Recognition 50

IX. Blood Sample Preparation 53

X. Accuracy Evaluation 54

References 55

I. INTRODUCTION

This chapter discusses several aspects of automated leukocyte differential analyzers, including some history of their development, what they do, how they work, and how to evaluate their performance. Section II describes the leukocyte (white blood cell) differential count as it is performed in a clinical laboratory. Some justification for automating this high-volume laboratory test is included. Section III gives an overview of the development of automated analyzers from a historic view. Section IV briefly discusses the major attributes of histochemical and pattern recognition machines. This chapter will be devoted almost entirely to the pattern recognition machines. Section V opens the topic of automation by pattern recognition techniques. The various features of an automated analyzer are discussed, along with the tasks it must be capable of performing. Section VI deals more directly with the hardware required to implement an automated analyzer. The hardware is described at a subsystem level, including performance requirements. Section VII describes with more rigor how some computerized image processing techniques can be used to extract the attributes of a leukocyte from a visual image. The section provides some insight into the techniques used in processing visual images with a computer. Section VIII completes the process started in the previous section by describing some computerized pattern recognition algorithms. These algorithms are used to compare the attributes extracted for a leukocyte sample with the attributes of leukocytes it is trained to recognize. Section IX deals with the topics of automated sample (slide) preparation and the desirability of producing samples of uniform quality. Section X discusses some recommended techniques for evaluating the performance of an automated analyzer.

The information presented in this chapter is a collection from the references given. The author's interpretation is based on his experience as a member of the team which developed the Corning Glass Works LARC® Analyzer. The techniques described are intended to provide some insight about the workings of any analyzer rather than describe a particular analyzer.

II. LEUKOCYTE DIFFERENTIAL COUNT

The effectiveness of the leukocyte (white blood cell) differential count as an indicator of an individual's health is well documented.[1,2] The test detects such health problems as inflammations (increased band neutrophils), allergic reactions (increased eosinophils), and leukemias (increased juvenile cells). The count is routinely performed; more than 240,000 are done each day in the U.S.[3]

The standard technique for performing the leukocyte differential count is labor intensive. A few drops of blood are smeared on a microscope slide and stained with a Romanowsky-type stain. The slide is examined with a microscope, usually at a magnification × 1000 with an oil-immersion objective. The area of the slide containing a monolayer of cells with good morphology is located. This area is scanned for leukocytes which are classified and counted. Encumbrances in the procedure include the 600 to 1 ratio of erythrocytes (red blood cells) to leukocytes and the shallow depth of focus at × 1000.

Each leukocyte is classified as one of the six normal types or as one of the many abnormal types, an abnormal type being any cell type not normally found in the peripheral blood of a healthy individual. Classification is based on such cell attributes as

1. Nuclear size
2. Nuclear texture
3. Nuclear optical density

Table 1
SUMMARY OF SOME ATTRIBUTES FOR THE NORMAL LEUKOCYTE CLASSES

Cell	Approximate nuclear size (μ^2)	Nuclear shape	Approximate cytoplasm size (μ^2)	Cytoplasm color
Segmented neutrophil	50	Two or more lobes	85	Pink, with fine granules
Band neutrophil	58	Band shaped, one lobe	85	Pink, with fine granules
Lymphocyte	60	Round to oval	30	Clear blue
Monocyte	105	Indented, may have holes	95	Gray to blue, with fine granules
Eosinophil	65	One or more lobes like the neutrophils	80	Pink, with large orangish granules
Basophil	55	One or more lobes like the neutrophils, often obscured by the cytoplasm granules	85	Pink, with large dark granules

4. Nuclear shape
5. Cytoplasm size
6. Cytoplasm texture
7. Cytoplasm optical density
8. Cytoplasm color
9. Existence of special structures

Some weighting may be given to nuclear color. Table 1 summarizes the attributes of several leukocyte classes. Figure 1 is a sample leukocyte. For more details, the reader is referred to a text on the subject.[2]

The laboratory report usually contains a count of 100 leukocytes by class plus qualitative erythrocyte and thrombocyte (platelet) data. The erythrocyte data lists abnormal shapes, sizes, colors, and inclusions, with modifiers for the degree or quantity. The thrombocyte data is a judgment of the sufficiency and a report of abnormal forms.

There are several reasons why errors occur in the leukocyte differential count. A good technologist, working at peak efficiency, takes 5 min to complete each blood sample. The visual examination is fatiguing; hence, the technologist's performance degrades as the day continues. This results in judgment, counting, and transcription errors which degrade the accuracy. In addition, the monotony makes it difficult to keep well-trained laboratory personnel.

Because the microscopic examination of blood smears is a high-volume, labor-intensive test, many researchers view it as an excellent candidate for automation. Advances in digital image processing, pattern recognition, and computer technology provide the means for automating the visual examination. As a result, several automated leukocyte differential analyzers are commercially available.

III. HISTORICAL BACKGROUND

In the mid-1960s, reports on automated techniques were published by researchers, such as Ingram and Preston[4] and Prewitt and Mendelsohn.[5] Reports from the research activities increased between 1968 and 1970. Ingram and Preston[1,6] published several

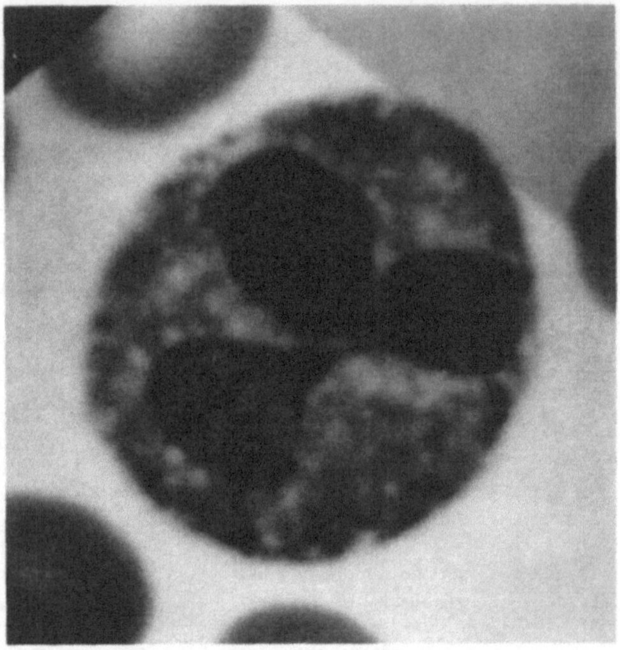

FIGURE 1. Sample leukocyte, segmented neutrophil.

more papers. Their paper "Automatic Analysis of Blood Cells"[1] described a prototype machine using image processing transforms defined by Golay.[7] Two Ph.D dissertations were written on the subject by Young[8] and Bacus.[9]

An important part of Bacus' work was his diligent effort to obtain statistically significant accuracy measurements for his automatic classifier and human observers. To this end, he collected a data base containing 1041 leukocyte images. The data base was divided into a 523-cell training set for developing the automatic classifier and an independent 518-cell testing set. The true classification for each leukocyte in the data base was obtained from the consensus opinion of qualified observers. The automatic classifier and the observers were compared to the consensus in order to develop conclusions about the accuracy of the automatic classifier relative to human observers.

The work mentioned thus far describes automated analyzers based on pattern recognition techniques. These methods automate the microscope and observer. Some researchers took a completely different approach, developing histochemical techniques to selectively stain the leukocyte classes for flow system counting.[10] This technique is used in the Technicon Hemalog-D® — the first commercially introduced instrument.

The first commercial automated differential analyzers based on pattern recognition were the Corning Glass Works LARC®[11,12] and the Geometric Data (Smith Kline) Hematrak.®[13] Both were introduced in October 1972 at the American Society of Clinical Pathologists Conference in San Francisco. However, it was 1974 before versions of these instruments reached the marketplace in quantity.

The techniques used in LARC® were developed from Bacus' work. However, many man years of research by Corning scientists[12,14-16] were required to refine the system.

Since the introductions of LARC® and Hematrak,® other companies have announced automated leukocyte differential analyzers, all using pattern recognition. Three of these analyzers are the Coulter Differential® (Cognos), the Perkin-Elmer diff3®, and the Abbott ADC 500®. Each of these instruments attempts to improve its performance over its predecessors by reporting more classes, sampling more cells, and automating a larger portion of the microscopic examination. A few of the newer designs attempt to classify some of the abnormal cell types and automate the erythrocyte and platelet reports.

IV. PATTERN RECOGNITION VS. HISTOCHEMICAL TECHNIQUES

The major advantage of the histochemical approach is its speed; thousands of cells can be classified and counted in 1 min compared to hundreds for the pattern recognition machines. This results in a more statistically significant differential as some types occur in a ratio of 1 or less per 100 leukocytes. The major disadvantage is the inability of current systems to differentiate enough classes, requiring a microscopic examination of the samples reported abnormal by the analyzer. Furthermore, a slide is still required to obtain the erythrocyte and thrombocyte data.

The pattern recognition machines have the advantage of melding with the practices used in the manual procedure. The classification criterion is based on the morphological features used in the manual procedure, and the slide is available for other tests. A microscope viewing port is provided to allow the operator to review suspect cells identified by the instrument and to make thrombocyte and erythrocyte observations during the procedure. Furthermore, a quality control protocol using the operator's and the instrument's classification of the same cells can be provided. The disadvantage is the slower rate for processing cells, resulting in fewer cells being counted. The argument against this disadvantage is that the results are at least as good as standard manual practices.

The pattern recognition instruments will be the primary subject of the remaining sections as the commercial units have tended to be based on this technology, and the majority of the published work has been dedicated to them.

V. AUTOMATION

The microscopic examination of a blood sample can be divided into several subtasks:

1. Preparing the blood smear on a microscope slide
2. Staining the smear
3. Loading the slide into the microscope
4. Focusing the microscope
5. Finding leukocytes
6. Classifying and counting 100 or more leukocytes
7. Examining and reporting erythrocyte morphology
8. Estimating thrombocyte sufficiency
9. Transcribing the results to report forms

Which subtasks should be automated? The "super" machine would automate all nine. Whole blood could be loaded into tubes in a magazine and identified. The results would be recorded on a report form, probably including total leukocyte, erythrocyte, and thrombocyte counts per cubic millimeter of whole blood.

The super machine was too magnanimous a task for the first instruments. Besides the material-handling problems, the state-of-the-art classification algorithms and hardware did not provide a good balance between accuracy, speed, and cost. Operator interaction had to be designed into the system.

The early analyzers automate Steps 4, 5, 6, and 9. A microscope viewing port and a keyboard are provided for the operator to observe and record erythrocyte and thrombocyte data. The classifier only solves the seven-class problem, classifying each cell as one of the six normals or as an abnormal; therefore a stop-on-unknown or review procedure is provided to allow operator identification of suspect cells. The analyzer-operator team can process about 40 slides per hour with this division of tasks.

The review procedure has an advantage over the stop-on-unknown procedure, as it allows the analyzer to complete its automated tasks unattended. The operator is required only at the end to identify unknown cells whose stage coordinates were stored during the automatic scan. This allows the technologist time to prepare other samples while the machine completes its tasks. Newer designs which process slides from a magazine review unknown cells after all slides in the magazine have been scanned.

Future trends are to automate more of the subtasks. The newer instruments provide some degree of automation for Steps 3 through 9. They have extended classifiers which identify more cell classes, but still require operator interaction to identify cells which cannot be identified with a high degree of certainty.

The hardware and software (algorithm) requirements for automating the various subtasks will be explored in subsequent sections.

VI. HARDWARE SUBSYSTEMS OF AN AUTOMATED LEUKOCYTE DIFFERENTIAL ANALYZER

Although the exact procedures may vary, the current analyzers perform several common functions:

1. Locating cells
2. Maintaining focus
3. Digitizing cell images
4. Processing images to find each cell
5. Extracting attributes for each cell
6. Classifying each cell
7. Counting cells by type
8. Reporting results

In order to perform these functions, an analyzer contains these or equivalent subsystems:

1. A microscope with motor driven x-y stage and focus mechanisms
2. A focus servo system
3. An acquisition processor which directs the microscope stage to bring potential cells into a window for recognition
4. An image scanning subsystem to extract a representation of the visual data for the image processing subsystem
5. An image processing and pattern recognition computer or computers
6. Input interfaces for operator supplied commands and data
7. Microscope viewing ports, such as binoculars and video displays
8. Output interfaces such as printers and displays for reporting results

A block diagram of such a system is in Figure 2.

The microscope subsystem contains many standard optical microscope components plus some additional optical elements for coupling other subsystems into the light path. Precise electromechanical drives are connected to the x-y stage and focus mechanisms. The stage may move in 5-μm or smaller steps or in several hundred-micron steps, depending on the cell acquisition scheme employed. The focus mechanism requires a resolution of about 1/10 μm, as the microscope usually employs a high-magnification,

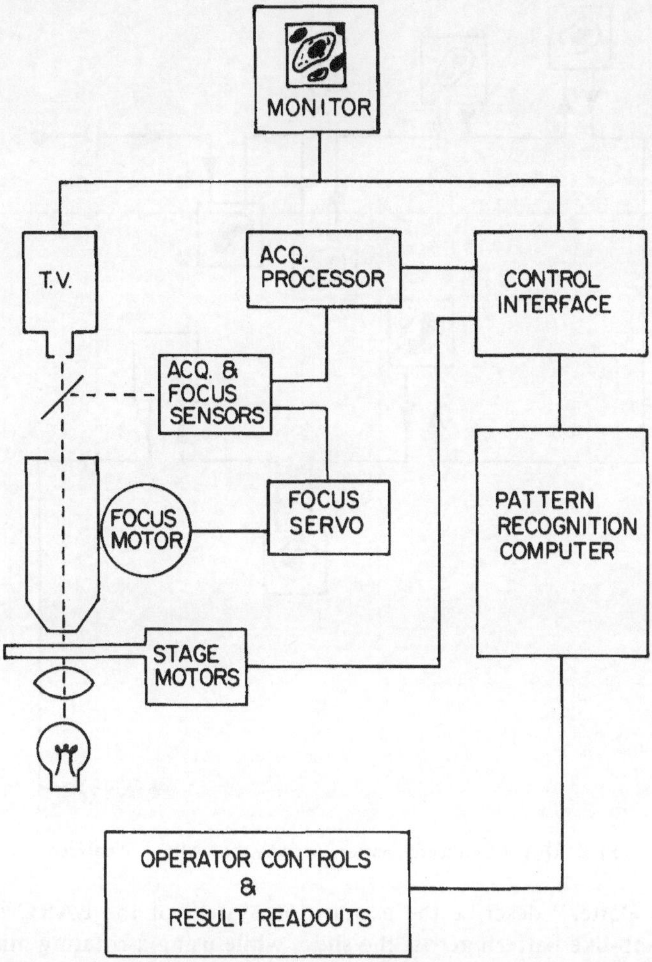

FIGURE 2. Block diagram of an automated leukocyte differential analyzer.

oil-immersion objective which has a shallow depth of focus. More importantly, the subsystem is designed so as to eliminate mechanical vibrations in the electromechanical drives as these vibrations tend to be magnified by the optics and appear as noise in the data extracted by the focus, acquisition, and image scanning subsystems.

The focus subsystem is simple in concept, but not in implementation. The focus servo seeks to maximize the derivative of the signal from an image scanning subsystem, thus, enhancing the cell edges. An optical filter may be used to enhance the erythrocyte images, the most populous blood component, improving the signal-to-noise ratio of the feedback signal in the focus servo. Because the depth of focus is only a few tenths of a micron, care must be exercised in designing the automated microscope. Vibrations coupled from the focus drive, x-y stage drive, or other sources can cause major problems in the focus servo system.

The acquisition subsystem locates leukocytes and neucleated erythrocytes and presents them to the image scanning subsystem for processing. The acquisition subsystem usually takes advantage of the high optical density of the cell nucleus to locate potential leukocytes and neucleated erythrocytes. The discrimination of the nucleus may be enhanced by using an optical filter that encompasses the wavelength of peak nuclear absorption.

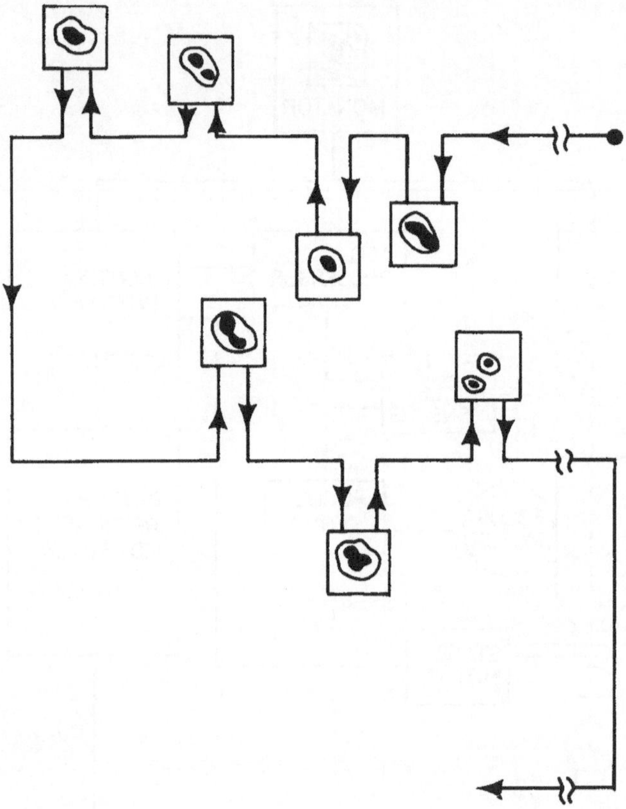

FIGURE 3. Cell search pattern used by the LARC® Analyzer.

Megla[11] and Cotter[12] describe the acquisition system of the LARC® Analyzer as following a comb-like pattern across the slide, while using a rotating mirror to search for cells in a narrow sweep orthogonal to the direction of stage motion. If a cell is encountered within the microscope field on either side of the stage path, the acquisition subsystem moves the stage to center the cell in a 20 × 20 μm video input window. After the image scanning subsystem digitizes and stores an image of the cell, the stage returns to its center path and continues its scan. This path is in Figure 3.

The LARC® approach is mechanically intensive. Another method is to move the stage in large steps and electronically do a gross scan of the microscope field (approximately 300 μm in diameter). When a cell is located, an electronic zoom is used to scan the cell using a smaller window with higher spatial resolution.

To prevent distorted differential counts, a major design task is to insure that the acquisition subsystem is free of any bias toward a cell class. Sources of bias can be the setting of the optical density threshold for locating nucleated cells or the step size of the acquisition scan. The threshold and step size also affect the speed at which cells can be processed. A high threshold will discriminate against cell classes with low optical densities, but a low threshold will slow down cell processing by stopping on erythrocyte fields. A larger step size may pass over small cells, while a small step size will result in a slower acquisition speed. The acquisition subsystem should be optimized for accuracy and speed.

The image-scanning subsystem quantifies the cell image located by the acquisition subsystem. To do this, it measures the optical density at predefined points on a grid. Each grid point is referred to as a picture element or pixel. The two most common grid patterns are in Figure 4. The hexagonal grid used by Golay[7] is preferred by many

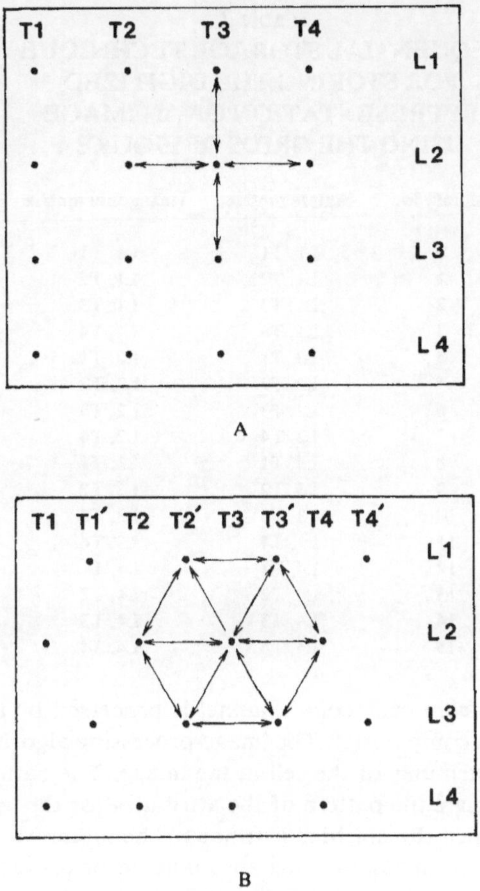

FIGURE 4. Image digitizing grid patterns. (A) Square grid, (B) hexagonal grid.

researchers because the distances between neighboring points are equal. The usual technique is to divide a microscope image which may represent 20 × 20 to 30 × 30 μm into a 50 × 50 or finer grid. The digitized values from the grid are stored in sequential memory locations of the image processor as in Table 2. Figure 5 is the digitized representation of a neutrophil with optical density values ranging from 0 for transparent to 63 for opaque.

To obtain color information, two or more digitized representations are taken of a cell image, using different wavelength optical filters. The filter wavelengths are usually chosen to correspond with absorption peaks in the Romanowsky stain or in blood components such as hemoglobin.

There are many techniques for digitizing an image. A common technique uses a conventional video camera. When using a video system, the row numbers L1, L2, etc. in Figure 4A correspond with horizontal lines, and the column numbers T1, T2, etc. correspond with time positions during a horizontal sweep. Although the time periods for Figure 4B appear complicated, the time periods for odd numbered rows T1′, T2′, etc. are the periods for even number rows T1, T2, etc. displaced one half of a time period. When digitizing with a square grid, the distance traversed by the horizontal sweep in one time interval must equal the horizontal line spacing. For the hexagonal grid, the distance traversed in one time interval must be $\sqrt{4h^2/3}$, where h is the horizontal line spacing. Design problems include reduction of geometric distortion and minimization of variations in optical density measurements across the grid.

Table 2
SEQUENTIAL STORAGE TECHNIQUE FOR STORING THE DIGITIZED REPRESENTATION OF AN IMAGE USING THE GRIDS OF FIGURE 4

Memory loc	Square matrix	Hexagonal matrix
0	L1, T1	L1, T1′
1	L1, T2	L1, T2′
2	L1, T3	L1, T3′
3	L1, T4	L1, T4′
4	L2, T1	L2, T1
5	L2, T2	L2, T2
6	L2, T3	L2, T3
7	L2, T4	L2, T4
8	L3, T1	L3, T1′
9	L3, T2	L3, T2′
10	L3, T3	L3, T3′
11	L3, T4	L3, T4′
12	L4, T1	L4, T1
13	L4, T2	L4, T2
14	L4, T3	L4, T3
15	L4, T4	L4, T4

The digitized image from each color channel is processed by the image processing and pattern recognition computer(s). The image-processing algorithms extract features which quantitate the attributes of the cell in the image. The pattern recognition algorithms attempt to associate the pattern of the attributes for the unknown cell with the patterns of cells for which the machine is trained. The unknown cell will be classified as the type it most closely matches or as an unknown or artifact if a close match is not found.

The image processing and pattern recognition algorithms might be implemented in a single general purpose digital computer or divided among several computers. There could be special purpose image processing hardware such as the Golay Logic Computer.[1] Regardless of the hardware implemetation, certain concepts are used. These concepts will be presented in the sections that follow.

VII. IMAGE PROCESSING AND FEATURE EXTRACTION

Computer algorithms which fall into the categories of image processing and feature extraction operate on the digitized images described in Section VI. These algorithms seek to remove noise, locate objects or regions of interest, and to quantitate the attributes of the objects. To develop an appreciation for these algorithms, some of the general concepts of image processing need to be explored.

One of the more basic image processing concepts is that of connectivity. If an individual pixel meets the requirements for being part of some region, which region does it belong to, is it a perimeter point, and how many regions are in the scene? These and similar questions are answered by connectivity algorithms.

The concept of connectivity may be stated: A pixel which passes all prescribed tests for being a member of a region is a member of a particular region if one or more of its neighbors are members of that region.

The prescribed tests can be optical density thresholds, color limits or corequirements with neighbors. The neighbors of the point (L2, T3) in Figure 4A are (L1,T3),(L2,T4),(L3,T3), and (L2,T2) at a distance of unity and (L1,T4), (L3,T4),

26	28	29	29	28	32	31	34	33	30	31	31	28	28	27	26	26	27	27	29	27	28	30	30	31	32	28	24	17	18
28	28	30	31	31	33	34	34	30	31	30	30	29	29	27	27	27	27	26	26	27	27	28	29	31	34	33	26	19	18
26	27	31	31	33	33	32	31	31	33	32	32	30	29	28	28	26	26	27	26	26	26	28	30	30	32	30	28	20	19
26	29	29	31	31	32	32	32	34	35	34	33	28	28	27	27	27	27	29	27	28	27	27	30	30	32	33	30	21	23
27	28	28	28	28	28	29	35	35	34	33	30	30	28	27	28	27	30	29	28	31	31	28	29	33	33	34	30	23	26
26	29	26	27	28	30	30	31	35	36	30	32	31	29	28	29	28	30	30	31	38	51	33	31	32	34	33	31	24	27
28	30	30	28	29	30	30	31	32	35	32	32	33	29	30	31	30	30	34	35	49	55	50	41	36	34	34	29	25	27
31	32	31	27	31	30	31	30	31	35	35	38	51	36	40	40	34	35	34	36	47	57	56	55	56	41	38	31	25	26
31	32	31	27	31	32	32	34	34	49	45	53	56	53	52	52	36	35	38	42	56	57	57	57	57	54	46	34	27	26
29	30	29	29	28	30	45	40	52	56	55	56	57	56	55	54	46	36	49	55	57	58	58	58	57	55	49	37	26	25
24	26	27	30	30	46	55	55	57	58	57	57	56	56	56	56	50	41	54	57	58	58	58	58	58	56	53	41	25	24
23	26	29	32	37	55	57	57	58	58	58	57	57	57	57	56	51	44	54	56	58	57	57	58	58	57	55	46	25	22
24	28	30	35	49	56	56	58	59	58	58	58	56	57	57	54	50	45	54	55	56	57	57	58	56	57	55	42	27	21
26	29	29	42	54	56	55	58	58	58	57	57	57	56	55	47	43	50	55	56	57	58	58	58	57	56	53	36	28	21
25	28	31	52	55	55	56	57	57	57	56	56	55	54	46	38	41	46	52	57	58	58	58	58	56	56	45	34	28	21
21	25	38	55	56	57	56	57	56	56	55	54	46	46	38	35	33	37	53	57	57	57	58	56	50	47	36	35	27	19
21	25	35	53	57	57	57	56	56	55	54	51	53	42	44	51	34	47	55	55	57	55	56	49	37	35	33	31	23	18
21	24	34	50	55	56	58	57	57	57	54	55	56	55	55	56	52	54	53	52	51	37	43	36	32	34	33	27	23	18
22	24	30	43	56	58	58	58	58	58	57	57	57	57	56	57	56	55	51	49	33	29	32	29	29	34	32	28	20	18
21	23	27	35	57	58	58	58	58	58	58	57	57	57	58	57	56	53	39	34	28	26	28	28	30	31	31	23	20	17
21	23	25	31	54	58	58	58	58	58	58	57	57	58	57	56	55	46	32	29	29	32	27	26	28	30	30	20	18	18
22	22	25	29	49	56	54	57	57	58	58	57	58	57	56	53	53	37	28	28	28	27	28	28	28	30	23	18	18	17
20	23	25	29	41	48	52	55	57	57	57	55	57	57	52	38	39	28	27	26	25	26	30	29	27	25	19	18	17	16
21	23	26	32	32	35	39	55	54	54	53	54	51	55	39	31	30	27	26	26	25	30	31	29	22	21	18	16	17	16
21	26	29	32	30	33	34	40	39	38	50	39	34	46	31	31	27	26	26	27	27	29	30	24	19	18	17	16	16	16
19	27	27	29	33	32	33	35	34	31	35	33	31	34	32	33	27	27	30	27	25	21	23	19	17	17	16	16	16	16
16	23	26	28	32	36	33	33	31	34	34	34	34	30	32	33	32	30	29	25	21	21	19	18	17	16	16	16	16	17
15	18	22	24	31	34	32	32	32	32	33	32	31	33	33	31	35	30	26	24	25	25	21	20	18	16	16	16	16	17
15	17	17	18	26	26	27	31	30	31	32	31	31	31	32	29	32	26	27	26	27	27	25	24	23	17	16	17	17	18
15	16	16	18	19	20	22	27	26	27	29	29	27	31	28	26	27	27	27	27	28	28	27	26	25	21	18	18	17	18

FIGURE 5. A portion of the digitized image of a neutrophil showing the cell nucleus.

(L3,T2), and (L1,T2) at a distance of $\sqrt{2}$. The neighbors of the point (L2,T3) in Figure 4B are (L2,T2), (L1,T2′), (L1,T3′), (L2,T4), (L3,T3′), and (L3,T2′).

Connectivity algorithms are described by several researchers.[7,17] A general connectivity algorithm is given in the flow chart of Figure 6. This algorithm is based on starting with a seed point which has passed some fundamental test to begin a region. The seed point is marked and used as a center point. Neighbors of the center point which pass the prescribed tests and have not been marked are marked and put in a list. When all neighbors of a center point have been tested, a new center point is selected from the list and its neighbors are tested. The process continues until all list points have been used as center points. Each pixel is accessed by its memory location in Table 2. Its neighbors are accessed by adding offsets to the location of the pixel. The offsets for a square grid are -1, $-n$, $+1$, and $+n$ for the neighbors at a unity distance and $-(n+1)$, $-(n-1)$, $+(n+1)$, and $+(n-1)$ for the neighbors at a $\sqrt{2}$ distance. For the hexagonal grid they are -1, $-(n+1)$, $-n$, $+1$, $+n$, and $+(n-1)$, when the center point is on even numbered rows, and -1, $-n$, $-(n-1)$, $+1$, $+(n+1)$, and $+n$ for center points on odd numbered rows. In both cases, n is the grid dimension. For the grids in Figure 4, n is four.

Scene segmentation is a difficult image processing problem. Consider the problem of finding the nucleus and cytoplasm of the cell in the digitized image of Figure 5.

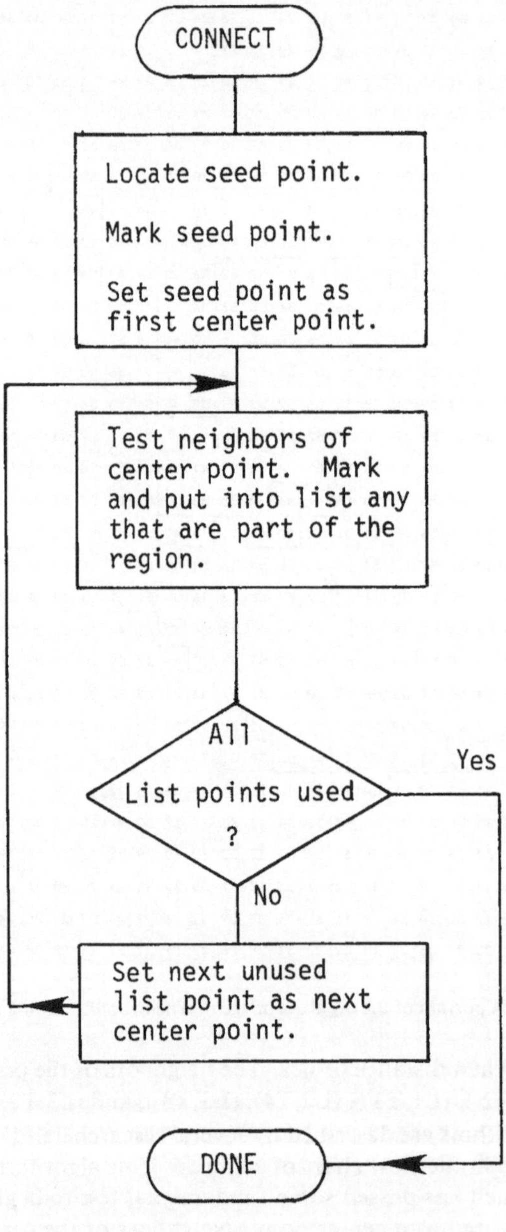

FIGURE 6. Flow chart for a connectivity algorithm.

The first step would probably be to set some optical density thresholds using a histogram as suggested by Cotter.[12] A histogram showing the distribution of optical densities for the image of Figure 5 is in Figure 7. The most apparent features of the histogram are the three major peaks which tend to encompass optical density values from the regions of interest. By using numerical smoothing techniques and taking numerical derivatives along the envelope of the histogram, thresholds can be set for the nuclear and cytoplasm regions.

One scene segmentation problem encountered is an overlap in the optical densities of the erythrocytes in the image and the cytoplasm of the leukocyte. Color information can help, but does not always offer a complete solution. Using a connection algorithm, the problem can be solved. First the nucleus is located and then the cytoplasm con-

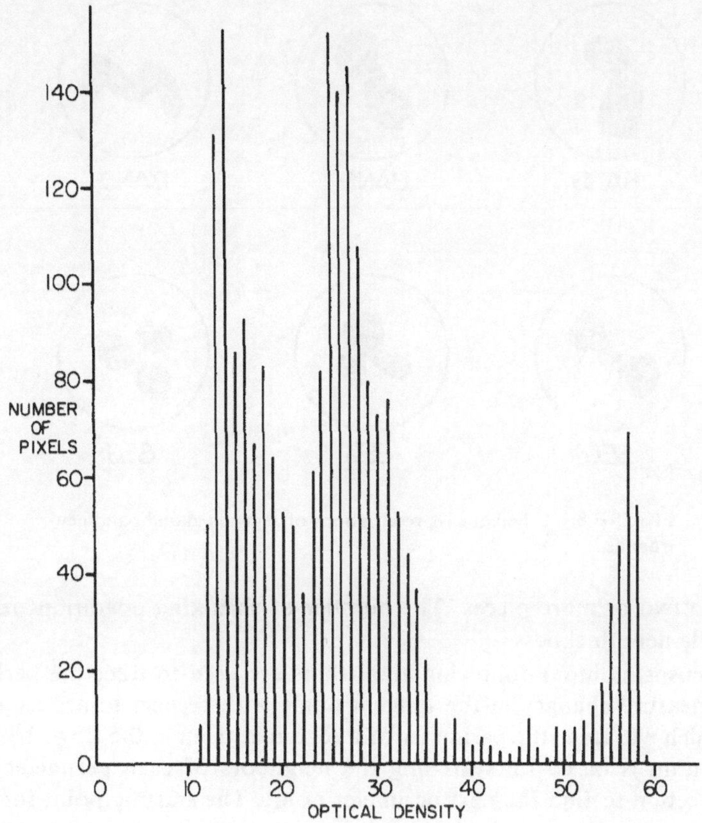

FIGURE 7. Optical density distribution histogram for the image in Figure 5.

nected to it. By using some fairly sophisticated algorithms to measure the degree of connectivity, erythrocytes which are touching the leukocyte can be eliminated. Thus, the leukocyte and its component parts are isolated.

As the cell and its parts are located, quantitative features can be extracted. The sizes of the nucleus and cytoplasm can be measured by counting the number of pixels in each. The optical density and color of each can be quantitated by average optical densities at several wavelengths. However, quantitative shape and texture features are more difficult to define.

The most common shape parameter is the length of the perimeter, squared and divided by the enclosed area, P^2/A. This feature is attractive because it is easily calculated and is a fair differentiator of round from more complex shapes. The area is measured by counting the pixels within the region and the perimeter by counting the pixels which belong to the region and have neighbors which do not. The minimum value of P^2/A is 4π for a circle, increasing as the shape becomes more complex.

More precise shape features are required to differentiate a segmented neutrophil from a band neutrophil.[16] Schematic examples of segmented and band neutrophils are in Figure 8. To differentiate the segmented neutrophil, attributes such as separate nuclear lobes connected by a fine filament and sharp identures or cusps in the nuclear perimeter must be detected. The latter assumes that cusps are the result of overlapping lobes when the cell is viewed in two dimensions.

Golay's algorithms[7] suggest a procedure for detecting and measuring filaments. It involves deleting perimeter points from the region and detecting when and if the region

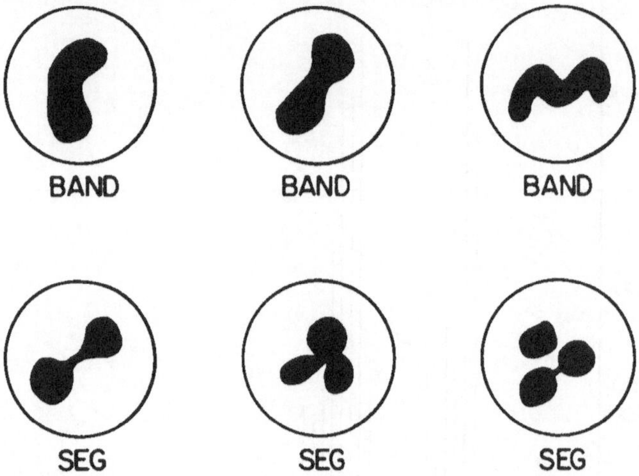

FIGURE 8. Schematic representations of segmented and band neutrophils.

segments into two or more pieces. The number of shrinking operations required is a measure of filament thickness.

To locate cusps, contour-following algorithms are used to trace the perimeter of a region and measure changes in the tangent angle with respect to arc length.[18-20] An algorithm which will trace the perimeter of a region is given in the flow chart of Figure 9. The algorithm is based on searching the neighbors of each perimeter point in a clockwise direction to find the next perimeter point. The starting point for testing the neighbors of each perimeter point is the last outside neighbor of the previous perimeter point. The perimeter is closed when the algorithm returns to its starting point and would leave at its starting angle.

Curvature is defined as the change in tangent angle with respect to arc length, $d\theta/ds$. The tangent is approximated by the angle of the vector to the next perimeter point found by the contour following algorithm. Continuous values are assigned to the tangent angle as a function of arc length. An increasing angle changes from $7\pi/4$ to $9\pi/4$, not $\pi/4$; likewise a decreasing angle changes from $\pi/4$ to $-\pi/4$, not $7\pi/4$. An estimate of the curvature is obtained from the tangent angle function by numerical differentiation techniques. If the direction of increasing angle is defined in a counter-clockwise direction, cusps will have positive curvatures. For the Figure 9 algorithm, cusps are negative.

Texture features are used to quantitate the visual texture of the cell's nucleus and cytoplasm. If the nucleus (cytoplasm) has a granular appearance, then its optical density has high spatial frequency components. The spatial frequencies at which the optical density varies can be measured with Fourier transforms,[21] however, the transform algorithms put a heavy computational burden on the image processing computer. To decrease the computational load, an approach which is less rigorous mathematically may be used. These algorithms may count the dark-to-light transitions across the nucleus (cytoplasm) or may look at the variance in the optical density between each pixel or group of pixels and its neighbors. Granules may even be sized and counted using connectivity algorithms. Whatever the algorithm, the resulting feature(s) should represent the degree of granularity. The major problem is to develop computationally efficient algorithms which do not degrade the separation between cell classes with noise. The noise in the value of the texture feature is often the result of natural transitions in optical density which the algorithm misinterprets as a granular texture. To

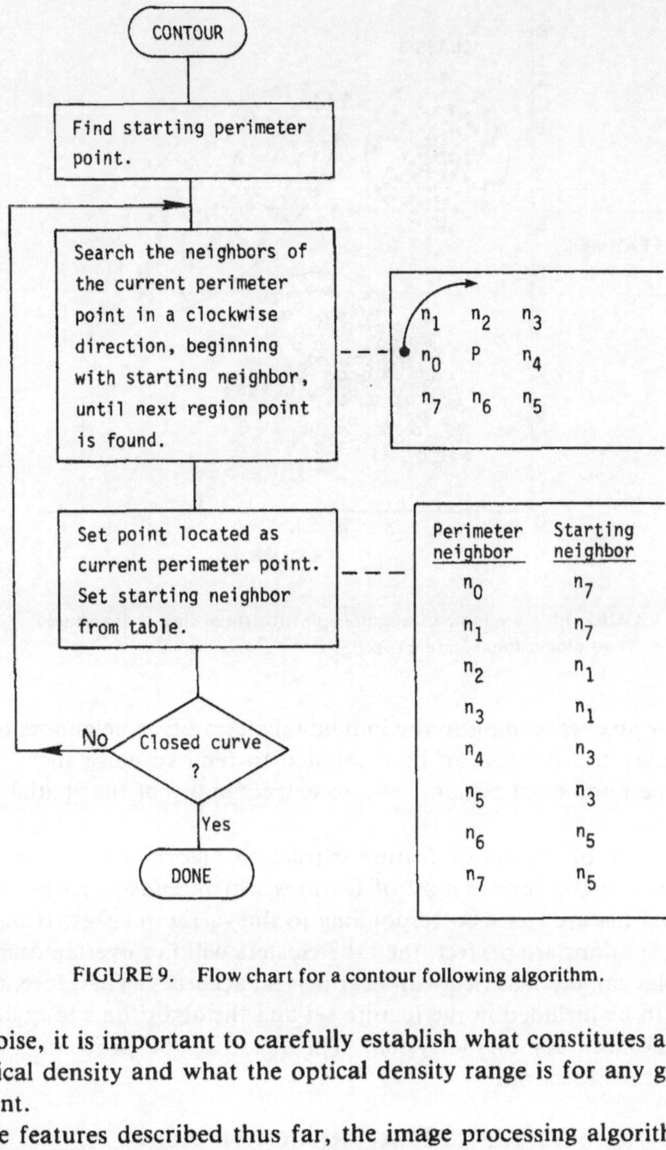

FIGURE 9. Flow chart for a contour following algorithm.

reduce the noise, it is important to carefully establish what constitutes a genuine transition in optical density and what the optical density range is for any granules which may be present.

Besides the features described thus far, the image processing algorithms may look for special structural details such as nucleoli within the nucleus. Nucleoli appear as small, round, or oval structures 2 to 4 μm in diameter. Their optical density is fairly homogeneous, and their color is about the same as that of the cytoplasm. These bits of cytoplasmic material are an indication of immaturity in a cell; therefore a nucleoli feature can enhance an instrument's ability to recognize immature cell types such as myeloblasts. Nucleoli can be located and identified, if they are present, with a connectivity algorithm. The algorithm must search for clusters of pixels within the nucleus that have the proper optical density, color, size, and shape. The resulting feature may be a count of the nucleoli.

The use of Golay transforms in automated leukocyte differential analyzers can speed the image processing algorithms.[1,7] The forte of these transforms is the ease with which they may be implemented in special purpose image processing computers which are capable of performing several operations in parallel. Golay transforms are defined for hexagonal image scanning grids of the type in Figure 4B. The transforms operate on binary coded images where a 1 indicates that a pixel is part of a region and a 0 indicates that it is not. Each Golay transform defines a set of conditions which modify the

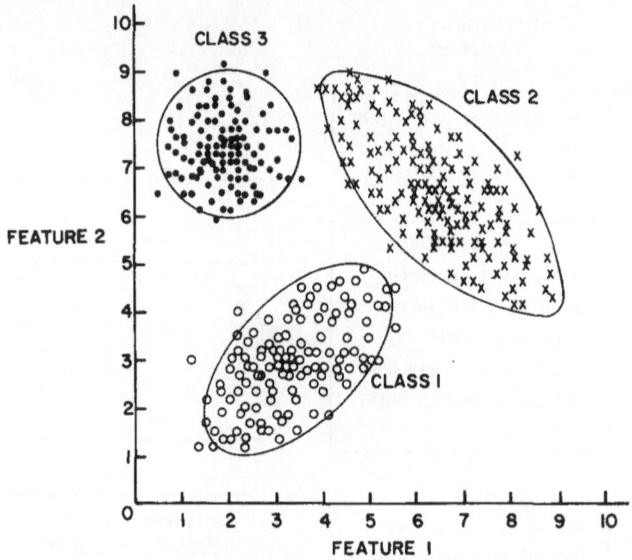

FIGURE 10. Example of a scatter plot for three classes distributed in a two-dimensional feature space.

binary state of a pixel according to the individual states of its neighbors or surround. The various Golay transforms can be combined to remove noise such as holes in a region, count the number of regions, and to extract many of the spatial features described previously.

The ultimate task of the set of feature extraction algorithms used in any pattern recognition system is to produce a set of features which tends to form clusters in the multidimensional feature space corresponding to the various classes. If the feature set and the class definitions are perfect, then the clusters will not overlap each other, and unknown samples can be classified with near perfect accuracy. Therefore, the selection of the features to be included in the feature set and the algorithms to quantitate them greatly affect the accuracy of the system. The classification problem is discussed in more detail in the next section.

VIII. PATTERN RECOGNITION

The list of quantitative attributes produced by the feature extraction algorithms, often referred to as a feature vector or pattern vector, is processed by a pattern recognition algorithm to arrive at a classification for the unknown cell. Pattern recognition involves the application of a decision rule which is developed during a training procedure using labeled samples of the classes to be identified.

A decision rule for a two-feature problem can be visualized with the aid of a scatter plot as in Figure 10. This plot graphs the distribution of labeled feature combinations for three classes. Decision boundaries may be drawn around each cluster as shown. If the feature vector for an unknown falls within one of these boundaries, it will be identified as a member of the corresponding class. Unknowns whose feature vectors fall outside the boundaries for all classes may be rejected as artifacts or brought to the operator's attention.

If there are more than two or three elements in the feature vector, it is impossible to visualize decision boundaries. Therefore, algorithms are used which allow a computer to develop functions that establish decision boundaries in a multidimensional feature space.

Decision functions are often defined with conditional probability density functions, $P(\bar{x}/i)$, the probability of getting feature vector $\bar{x}$ given class i;[22] i.e., $P(\bar{x}/i)$ is the probability that a member of class i could be described by the feature vector $\bar{x}$. In Figure 10, $P(\bar{x}/1)$ is greatest when $\bar{x}$ is about (3.5,3.0) i.e., $\bar{x}$ has the highest probability of belonging to Class 1 if it is located in the region of feature space where the largest number of Class 1 samples cluster. For Classes 2 and 3, the $\bar{x}$s which maximize $P(\bar{x}/i)$ are (6.5,6.5) and (2.0,7.5), respectively.

It can be shown that a good statistical decision rule for assigning a classification, $C(\bar{x})$, is

$$C(\bar{x}) = k$$

if

$$L(j,k)\, P(k)\, P(\bar{x} \mid k) \geq L(k,j)\, P(j)\, P(\bar{x} \mid j)$$

for all

$$j \neq k \qquad (1)$$

where L(c,i) is the cost associated with choosing Class c when the true class is i, and P(i) is the *a priori* probability of having a sample from Class i.[22] Weighting $P(\bar{x}/i)$ by both the cost of classifying samples of Class i as other classes and the probability of having a sample from Class i minimizes the risk of making unacceptable classification errors.

In some cases, the expected fraction of Class i in the total population is used for P(i) in Equation 1. Doing this will minimize the percentage of classification errors. However, in a leukocyte differential analyzer, each blood sample must be considered independently. The population of certain cell types such as the juvenile classes may be low for the total population, but it can be high for any given blood sample. Since the analyzer has no *a priori* knowledge about each blood sample, it is usually best to use equal P(i)s for all classes. The L(c,i)s can then be fine tuned in an iterative manner to optimize the decision rule.

A procedure for obtaining $P(\bar{x}/i)$ in a two-dimensional feature space is to subdivide the space and count the number of samples of Class i in each subdivision. This is referred to as a histogram. The probability of a sample occurring in a subdivision is equal to the number of samples in the subdivision divided by the total number of samples in Class i. The probability density within each subdivision is the probability divided by the area of the subdivision. The true probability density is approached as the total number of samples approaches infinity and the area of the subdivisions approaches zero. The histogram approach may be extrapolated to higher dimensions, but is usually impractical for high-dimensional spaces because of the requirements for many subdivisions and many sample points.

Another approach is to assume a parametric form for the function and determine the parameters which will best fit the function to the sample data. The Gaussian or normal distribution is often used. For the multivariate case, it has the form[22]

$$P(\bar{x} \mid i) = \frac{1}{(2\pi)^{n/2} |Q_i|^{1/2}} \exp\left[-\tfrac{1}{2} (\bar{x} - \bar{m}_i)\, Q_i^{-1}\, (\bar{x} - \bar{m}_2)^{T}\right] \qquad (2)$$

where Q_i is the n-by-n covariance matrix for Class i, n is the number of elements in the feature vector $\overline{x}$, and $\overline{m}_i$ is the mean feature vector for Class i. The term $(\overline{x}-\overline{m}_i)^T$ is the transpose of the row vector $(\overline{x}-\overline{m}_i)$. $|Q_i|$ is the determinant of Q_i and Q_i^{-1} is the inverse.

The parameters $\overline{m}_i$ and Q_i which define the function are easily derived from a set of labeled samples. The mean feature vector for each class is

$$\overline{m} = \frac{1}{N} \sum_{j=1}^{N} \overline{y}_j \tag{3}$$

where $\overline{Y}_j$ is the feature vector for the jth sample of the class, and N is the number of samples. The terms of the covariance matrix for each class are

$$q_{kl} = \frac{1}{N} \sum_{j=1}^{N} (y_{jk} - m_k)(y_{jl} - m_l) \tag{4}$$

where q is the element of Q in the kth row and the lth column, y_{jk} and y_{jl} are the kth and l^{th} elements of the jth feature vector, and m_k and m_l are the kth and l^{th} elements of the mean feature vector for the class. A minimum of 2η samples of each class are needed to determine each $\overline{m}$ and Q, where η is the dimension of the feature space. For best results, several times this number should be used.

By finding the parameters of Equation 2 and establishing the costs and *a priori* probabilities in Equation 1 for each class, a decision function is completely specified.

There is little reason for not using a Gaussian model if the distribution of feature vectors for each class is essentially Gaussian or if the model provides an acceptable degree of accuracy. The Gaussian model is easily implemented, requiring little storage for parameters and a simple computer program, compared to many other models. When the Gaussian model will not work, other techniques can be used to estimate P $(\overline{x}/i)$. These techniques may use potential functions[22] or other procedures to estimate P $(\overline{y}_j/i)$ at each sample point $\overline{y}_j$ and then apply surface-fitting techniques to find a functional description for P $(\overline{x}/i)$. The function may be a piecewise linear approximation or a higher order approximation. Once a procedure for finding P $(\overline{x}/i)$ is found, Equation 1 can be used for the decision rule.

Rather than applying Equation 1 to all attributes simultaneously, a pattern recognition algorithm may be organized in a tree structure as in Figure 11. The nodes represent classifications and the branches decisions. The decision rules for the branches may use single attributes or combinations of several. For example, the first branches might separate cells with blue cytoplasm from those with red, while other branches might use a combination of nuclear size and shape to further separate classes. The decision rules may be based on Equations 1 and 2 or other functions which minimize the risk of errors.[22]

A tree structure approach is especially useful when an attribute adds superfluous information to the recognition procedure for some classes. This superfluous information may actually degrade the discrimination between classes which do not require it. Examples are the features, used primarily to differentiate segmented from band neutrophils, which count nuclear lobes and cusps in the nuclear perimeter. A pattern recognition algorithm would first determine that a leukocyte was a neutrophil and then use the lobe and cusp features to discriminate band neutrophils from segmented neutrophils.

When the image processing, feature extraction, and pattern recognition algorithms are merged, a system for recognizing visual images of blood cells is completely speci-

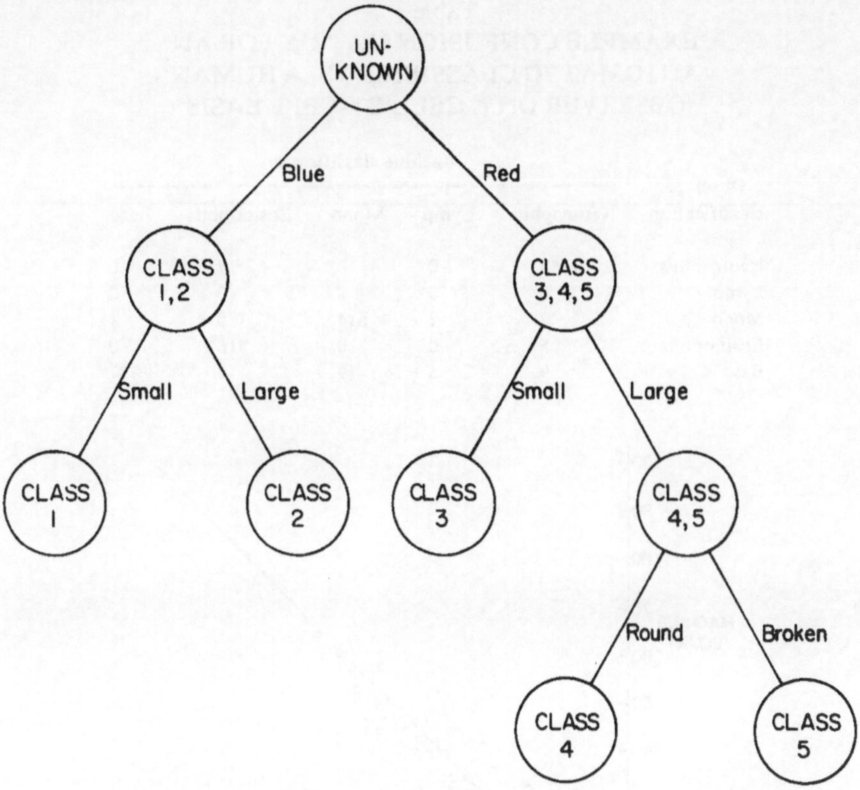

FIGURE 11. Example of a tree structure decision rule used for pattern recognition.

fied. These algorithms are not developed independently of each other, but are refined together to optimize the blood cell recognition accuracy. The feature extraction algorithms must provide sufficient information for the pattern recognition algorithms, and the pattern recognition algorithms must be designed to use the available information.

IX. BLOOD SAMPLE PREPARATION

The quality of the blood sample as prepared on the microscope slide affects the accuracy of the automated leukocyte analyzer. The analyzer's pattern recognition algorithms are trained to expect a specific distribution of attribute values for each cell class. If the sample preparation process varies and the distribution of an attribute shifts, then the probability of classification errors increases.

The need to reduce the effect of sample preparation variables on the pattern recognition algorithms and reduce the sample preparation time prompted several researchers to consider techniques for automating the procedure. Megla[11] and Rogers[14] described the LARC® Spinner for spreading the blood sample in a monolayer on a microscope slide. The LARC® Stainer which applies stain to the prepared slide and controls the staining reaction was described by Megla.[11] Other techniques were described by Ingram and Minter[24] and Preston and Norgen.[25]

The chemistry and quality control of the stain components are very important if reproducible staining is to be obtained from one stain batch to the next. Dean and associates used spectrophotometry, liquid chromatography, and a modified version of the LARC® Analyzer to quantitate degradations of blood stains.[15] Using high-performance liquid chromatography, they were able to resolve the thiazine dyes compos-

Table 3
EXAMPLE CONFUSION MATRIX FOR AN AUTOMATED CLASSIFIER VS. A HUMAN OBSERVER ON A CELL-BY-CELL BASIS

Observer classification	Machine classification: Neutrophils	Lymp	Mono	Eosinophils	Baso
Neutrophils	1558	0	6	7	0
Lymp	0	850	8	0	0
Mono	3	2	614	0	1
Eosinophils	5	0	0	416	0
Baso	9	8	19	0	137

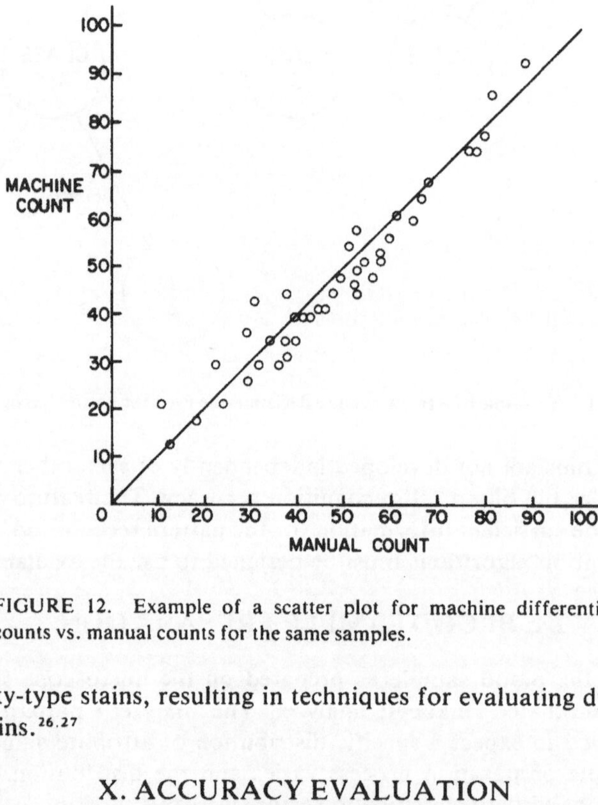

FIGURE 12. Example of a scatter plot for machine differential counts vs. manual counts for the same samples.

ing Romanowsky-type stains, resulting in techniques for evaluating dye purity and for constructing stains.[26,27]

X. ACCURACY EVALUATION

The accuracy of an automated leukocyte differential analyzer system can be evaluated at two levels. At the microscopic level, the cell-by-cell classification of the analyzer can be compared to those of an observer and the results tabulated in a confusion matrix as in Table 3. In Table 3, the columns are the machine's classifications and the rows are the observer's; therefore the entry in Row 1 and Column 4 indicates that the machine classified as eosinophils seven cells which the observer classified as neutrophils. At the macroscopic level, manual and machine differentials can be compared slide-by-slide and the results can be scatter plotted for each cell type as in Figure 12. Each point on Figure 12 represents the machine count vs. the manual count for one blood slide. Points below the diagonal represent machine counts which are lower than the manual, and those above represent machine counts which are higher.

To fully test an analyzer, the classification accuracy should be evaluated on a cell-by-cell basis, and the system performance should be verified on a slide-by-slide basis. On a cell-by-cell basis, the confusion matrix reveals potential problems such as classifier errors which would tend to cancel when processing more or less normal slides. On a slide-by-slide basis, scatter plots reveal the trend of the analyzer for each cell class such as a bias for a particular class. A bias could be attributed to a classifier problem which often confuses members of one cell class with those of another or a bias in the cell acquition subsystem.

Given the accuracy data, the most difficult evaluation task is the determination of the analyzer's acceptability. The minimum performance level is for the reported distribution of blood cell types to be within normal ranges for the majority of normal samples and for it to call attention to clinically abnormal samples. The ideal analyzer reports an essentially correct distribution for both normal and abnormal samples. A cost-effective analyzer probably has a performance level between these two extremes.

The most undesirable errors are those which contribute to false negative results, causing a sick individual to be diagnosed as healthy. However, an excessive number of false positive results degrades the machine's value by forcing the operator to manually verify positive results. This statement applies at the microscopic level when the machine requests assistance in classifying a cell and at the macroscopic level when the sample's diagnosis is determined from the reported differential count. Therefore, an analyzer may tolerate a number of specific errors to achieve a desired performance. For example, an analyzer may be tuned to classify some metamyelocytes as band neutrophils to prevent a large number of band neutrophils from being classified as abnormal cells. Compromises such as this can reduce false positives, but they must not allow clinically significant differential counts to pass unnoticed.

As a final note, when evaluating an analyzer, more precise conclusions are obtained if the machine is compared to a consensus classification obtained from several independent observers.[27] The expected accuracy for an analyzer compared to this consensus is about that of an average observer.

REFERENCES

1. **Ingram, M. and Preston, K., Jr.,** Automatic analysis of blood cells, *Sci. Am.*, 223, 72, 1970.
2. **Miale, J. B.,** *Laboratory Medicine: Hematology,* 4th ed., C. V. Mosby, St. Louis, 1972.
3. **Young, I. T.,** The classification of white blood cells, *IEEE Trans. Biomed. Eng.*, BME-19, 291, 1972.
4. **Ingram, M. and Preston, K., Jr.,** Importance of automated pattern recognition techniques in the early detection of altered blood-cell production, *Ann. N.Y. Acad. Sci.*, 113, 1066, 1964.
5. **Prewitt, J. M. S. and Mendelsohn, M. L.,** Analysis of cell images, *Ann. N. Y. Acad. Sci.*, 128, 1035, 1966.
6. **Ingram, M., Norgren, P., and Preston, K., Jr.,** Automatic differentiation of white blood cells, in *Image Processing in Biological Science: Proceedings of a Conference Held November, 1966,* Ramsey, D. M., Ed., University of California Press, Berkeley, 1968.
7. **Golay, M. J. E.,** Hexagonal parallel pattern transforms, *IEEE Trans. Comput.*, C-18, 733, 1969.
8. **Young, I. T.,** Automated Leukocyte Recognition, Ph.D. thesis, Massachusetts Institute of Technology, Cambridge, 1969.
9. **Bacus, J. W.,** An Automated Classification of the Peripheral Blood Leukocytes by Means of Digital Image Processing, Ph.D. thesis, University of Illinois, Chicago, 1970.
10. **Ansley, H. and Ornstein, L.,** Enzyme histochemistry and differential white cell counts on the Technicon Hemalog-D, in *Advances in Automated Analysis,* Technicon International Congress, Tarrytown, N. Y., 1970.
11. **Megla, G. K.,** The LARC automatic white blood cell analyzer, *Acta Cytol.*, 17(1), 3, 1973.
12. **Cotter, D. A.,** Automatic white blood cell differential analysis, *Am. J. Med. Technol.*, 39, 383, 1973.

13. **Miller, M. N.**, Design and Clinical Results of Hematrak — an Automated Differential Counter, Proc. 2nd Int. Joint Conf. Pattern Recognition, Lyngby-Copenhagen, 1974.
14. **Rogers, C. H.**, Blood sample preparation for automated differential systems, *Am. J. Med. Technol.*, 39, 435, 1973.
15. **Dean, W. W., Stastny, M., and Lubrano, G. J.**, The degradation of Romanowsky-type blood stains in methanol, *Stain Technol.*, 52, 35, 1977.
16. **Carter, C. N. and Cotter, D. A.**, Leukocyte nucleus shape evaluation — systems and methods, in *Proc. 6th Ann. Southeastern Symp. System Theory*, College of Engineering, Louisiana State University, Baton Rouge, 1974.
17. **Rosenfeld, A.**, Connectivity in digital pictures, *J. Assoc. Comput. Mach.*, 17, 146, 1970.
18. **Rintala, W. M. and Hsu, C. C.**, A feature-detection program for patterns with overlapping cells, *IEEE Trans. Syst. Sci. Cybernet.*, SSC-4, 16, 1968.
19. **Bennett, J. R. and MacDonald, J. S.**, On the measurement of curvature in a quantized environment, *IEEE Trans. Comput.*, C-24, 803, 1975.
20. **Freeman, H. and Davis, L. S.**, A corner-finding algorithm for chain-coded curves, *IEEE Trans. Comput.*, C-26, 297, 1977.
21. **Hildebrand, F. B.**, *Advanced Calculus for Applications*, Prentice-Hall, Englewood Cliffs, N. J., 1962, 236, 442.
22. **Meisel, W. S.**, *Computer-Oriented Approaches to Pattern Recognition*, Academic Press, New York, 1972.
23. **Ingram, M. and Minter, F. M.**, Semiautomatic preparation of coverglass blood smears using a centrifugal device, *Am. J. Clin. Pathol.*, 51, 214, 1969.
24. **Preston, K., Jr. and Norgen, P. E.**, Method of preparing blood smears, U.S. Patent 3,577,267, 1971.
25. **Dean, W. W., Lubrano, G. J., Heinsohn, H. G., and Stastny, M.**, The analysis of Romanowsky blood stains by high-performance liquid chromatography, *J. Chromatogr.*, 124, 287, 1976.
26. **Lubrano, G. J., Dean, W. W., Heinsohn, H. G., and Stastny, M.**, The analysis of some commercial dyes and Romanowsky stains by high-performance liquid chromatography, *Stain Technol.*, 52, 13, 1977.
27. **Bacus, J. W.**, The observer error in peripheral blood cell classification, *Am. J. Clin. Pathol.*, 59, 223, 1973.

COMPUTERS IN CLINICAL INSTRUMENTATION

Stephen C. Dunning

TABLE OF CONTENTS

I. Introduction 58

II. Digital Computer History 58

III. Digital Computer Fundamentals 58
- A. Numbers and Computers 58
- B. Classical Computer 59
- C. Basic Computer Operation 60
- D. Classes of Computers 61

IV. Applying Computers in Instruments 63
- A. Sequencing and Control 64
- B. Data Input 64
- C. Measurement Techniques 65
- D. Calibration 67
- E. Derived Results 67
- F. Simplification 68
- G. Some Limitations 68

V. Future Developments 68

VI. Conclusion 69

References 70

I. INTRODUCTION

The modern digital computer has had a dramatic impact on the capabilities of clinical instrumentation, both in new abilities to measure and in improved techniques. These advances should result in better medicine and in improved health care at lower cost to the public.

This chapter will first present a brief review of the history of the digital computer and develop some fundamental concepts, so that the uses, limitations, and future trends of digital computers in clinical instrumentation can be better understood.

II. DIGITAL COMPUTER HISTORY

The modern digital computer has deep roots, starting with the invention in 1642 of the first adding machine by the French philosopher and mathematician Blaise Pascal. The 19th century British mathematician Charles Babbage refined the concepts of the calculator and conceived of an "analytical engine" designed to solve complex mathematical problems. While the mechanical technology of his time prevented Babbage from building a working engine, his concepts of a calculating machine being controlled by a sequence of instructions (a stored program) are sound today.

The advances of electronics and the needs of war came together in the 1940s to produce several early electronic computers. These machines were large, expensive, and difficult to program, but the speed with which they solved important problems justified the efforts.

The architecture of the digital computer of today is based on the stored program concepts of John von Neumann.[1] This architecture, with many subtle variations, is basic to all modern digital computers. The invention of the transistor in 1948 at Bell Telephone Laboratories really started the computer age by making possible the collection of reliable electronic circuits into large computer systems. Previous efforts were often limited by the short life expectancy of the vacuum tubes being used. The years since have seen a relentless effort to design and build smaller, more reliable, and more capable computers at ever lower cost. Today entire digital computers, consisting of thousands of transistors, are available as semiconductor components suitable for inclusion in such products as clinical instruments.

III. DIGITAL COMPUTER FUNDAMENTALS

The digital computer is a machine that can perform simple operations on numbers according to a list of stored instructions called a program. The computer has become such a powerful and necessary modern tool because it can tirelessly perform many thousand such simple operations with near perfect accuracy. By combination these basic operations are strung together by the programmer to provide useful numerical calculations and complex control functions. For example, a computer can organize a large data base and sort out a useful subset, control the operation of a large power plant, or monitor quality on an assembly line.

A. Numbers and Computers

A computer operates on numbers, and most problems can be expressed or coded using numerical notation, and events can be assigned numerical values, so that calculation and control operations can be performed. Numbers are represented by means of a set of symbols. The universal system of numbers is the decimal system, using the symbols 0, 1, 2, 3, 4, 5, 6, 7, 8, and 9. A numerical value in the decimal system is expressed as:

$$N = 315 = 3 \times 10^2 + 1 \times 10^1 + 5 \times 10^0 \quad (1)$$

where the value of place is a power of the base number ten. A number may be expressed in other bases according to the form:

$$N = A_m b^m + A_{m-1} b^{m-1} + \cdots + A_1 b + A_0 \quad (2)$$

where A_m, A_{m-1}, ..., A_0 are symbols, and b is the base.[2] Table 1 shows a number expressed in several different numbers systems.

A particularly useful number system for use in computers is the base 2 system or binary number system, where the symbols 1 and 0 correspond to the presence and absence of electrical signals. Many useful electronic circuits are available to detect, store, and operate on these binary conditions. Decimal numbers can be converted to binary and processed by the computer and then be converted back to decimal for display of results. An individual base 2 number (0 or 1) is called a bit. Bits are assembled into words, and an eight-bit word is called a byte. These binary words are used by the computer to define both numbers (data) and instructions (program steps) within the electronic circuits of the computer.

B. Classical Computer

The classical computer consists of the four basic subsystems in Figure 1. The memory stores both data and instructions in binary words. A typical memory is organized as a group of unique "pigeon holes," each defined by its own address. These locations are accessed by providing an address (see Figure 2A), and data may be read from or written to an individual location. Memory is available in many forms, including various types of semiconductor memory. Semiconductor memory has the characteristic that its contents will be lost when power is removed. A method must be provided to write instruction words into read-write semiconductor memory (called RAM for random access memory). A type of semiconductor memory (called ROM for read only memory) is often used when such a loader device would be too expensive or inconvenient. ROM is loaded one time, either by the vendor or in some cases by the user (called PROM for programmable).

The arithmetic-logic unit (ALU) performs operations on the binary numbers at its inputs (see Figure 2B) and places the result on its output. A storage register called the accumulator may be present on one of the inputs as both a main data source and a place to store operation results. Operations can be arithmetic, such as addition and subtraction, or logical operations, such as AND, OR, and EXCLUSIVE OR.

The input/output section makes the computer a practical and useful tool by providing the means of entering data and conditions into the computer and obtaining results or control from the computer. Input might be connected to a keyboard and the output to a printer.

The control subsystem of a classical computer generates those signals needed to control and sequence the other subsystems through their operations. The control section includes the program counter, a pointer that contains the address of the memory word that is the next program instruction to be executed, and the instruction register which stores and interprets the present instruction word so that the proper control sequences can be generated. A special class of instructions are available that allow the contents of the program counter to be modified based on the results of previous program operations giving the computer the capability to alter its instruction execution path and to show simple decision making ability.

Table 1
NUMBERS IN VARIOUS BASE SYSTEMS

Base 10	327
Base 8	507
Base 5	2302
Base 3	110010
Base 2	101000111

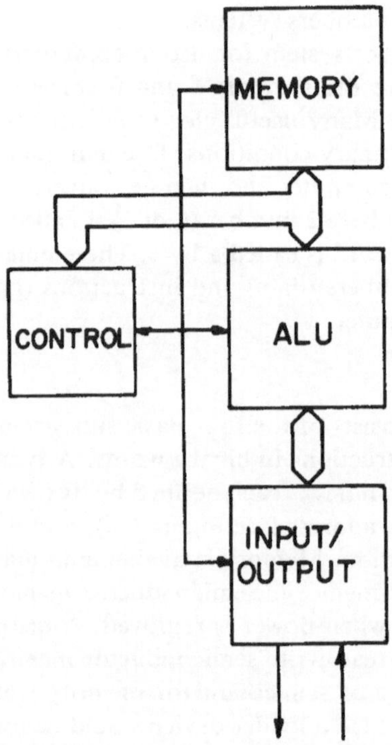

FIGURE 1. The classical computer model.

C. Basic Computer Operation

The operation of the basic subsystems of a digital computer can best be shown by an example. The operation of a simple computer-controlled "pill counting" machine will be used to illustrate calculation, control, and simple decision making. Figure 3 shows a motorized conveyer belt that transports pills to a point where they can fall into an open pill bottle. During the fall, each pill will activate a normally open switch. The rules of operation require that an operator place an open bottle in position under the belt and push a start button, at which time the machine will place eight pills in the bottle and stop. Figure 4 shows a flowchart representation of this sequence, and Table 2 is an example of a computer program to execute this flowchart. The pill detect switch is an input, and the motor control relay is an output. The program is shown in a descriptive form for example only and in practice would be written in a convenient mnemonic shorthand called assembly language. Using a computer program called an assembler, these mnemonics would be translated into binary words (machine language) to be stored in the computer memory.

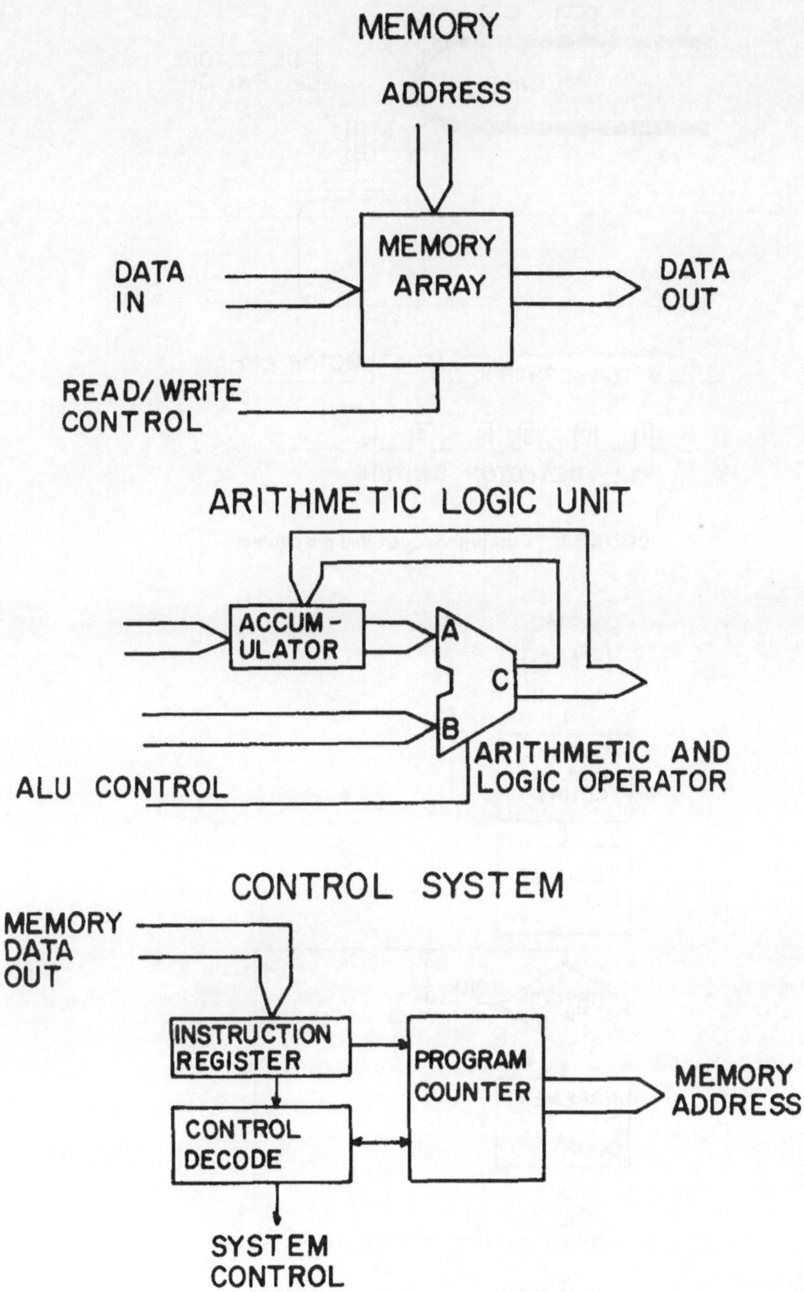

FIGURE 2. Computer subsystems.

D. Classes of Computers

Computers come in many shapes and sizes, depending on the tasks to be performed and the cost of the systems. Only a few years ago almost all digital computers were large and very expensive. To justify the expense of these systems, they were configured with many peripheral input-output devices and operated in a general purpose mode for many users. These large general-purpose machines still represent the largest amount of the installed value of present computers. They are found in banks maintaining deposit records, in large airline reservation systems, and in universities solving complex scientific problems.

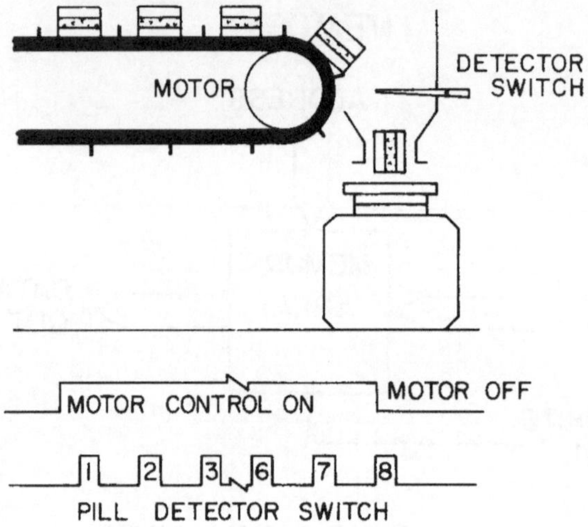

FIGURE 3. Computer-controlled pill counter.

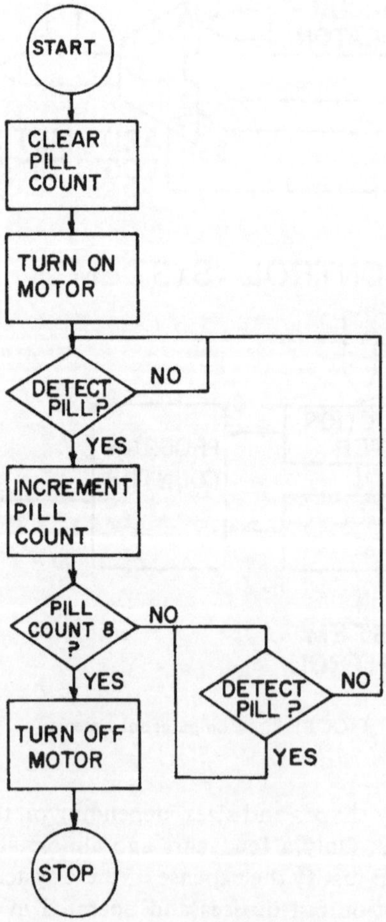

FIGURE 4. Flowchart for computer-controlled pill counter.

Table 2
SAMPLE COMPUTER PROGRAM

Label field	Instruction field	Comment field
START:	LOAD REGISTER B = 8	; PILL COUNT
	LOAD ACCUMULATOR = 01	
	OUTPUT TO MOTOR CONTROL	; TURN ON MOTOR
INSW1:	INPUT DETECT SWITCH	; CHECK FOR CLOSED
	BRANCH TO INSW1 ON ZERO INPUT	; SWITCH, CONTINUE
	DECREMENT REGISTER B	; COUNT PILL
	BRANCH TO STOP ON ZERO IN B	; STOP ON 8 PILLS
INSW2:	INPUT DETECT SWITCH	; CHECK FOR OPEN
	BRANCH TO INSW2 ON NOT ZERO	; SWITCH, CONTINUE
	BRANCH TO INSW1	; REPEAT
STOP:	CLEAR ACCUMULATOR	
	OUTPUT TO MOTOR CONTROL	; TURN OFF MOTOR
	HALT	

As technology improved, a smaller and less-expensive class of computers was developed. These computers (called minicomputers) allowed economical use of a dedicated processor in an industrial process or as a data monitor in a scientific experiment. Minicomputers are usually slower than the large general-purpose machines and can support less memory. Often minicomputers run in the "real-time mode," attached directly to the data source and exercising control over it. Technology advances have continued to lower the cost and increase the performance of the minicomputer, blurring the distinction between it and the general-purpose machines and creating a new class called the midicomputer.

The most-recent and most-dramatic development in digital computers is the availability of complete processor systems on just a few or even one large-scale integrated circuit. These components are called microprocessors, and the complete systems are called microcomputers. Costing only a few dollars, these devices offer the processing power that 10 years previously cost millions of dollars. They have turned the computer into a commodity, bringing the power and flexibility of processor-based solutions to the widest range of products. Microprocessors/microcomputers are found in toys, games, sewing machines, automobiles, medical instruments, function generators, and oscilloscopes, and the list can go on and on. These devices are the most significant development in modern electronics, and they are causing a revolution in the use of computers in everyday life. Figure 5 shows the organization of a typical single-component microcomputer.

All of these classes of computers are alike in that they all have the basic elements of the classical computer. A microcomputer can perform the same tasks that a large general-purpose computer can perform, but at a much slower rate. A combination of cost and performance determines the acceptability of a processor in a given situation. Technological advances continue to change these cost-performance relationships.

IV. APPLYING COMPUTERS IN INSTRUMENTS

A system is a prime candidate for a processor solution when it includes significant calculation or control or a combination of calculation and control. Most modern electronic instruments fall into this category, and a number of important functions can be accomplished with a processor, often with dramatic improvements in cost and performance.

The following sections outline the important application considerations when using a processor in an electronic instrument.

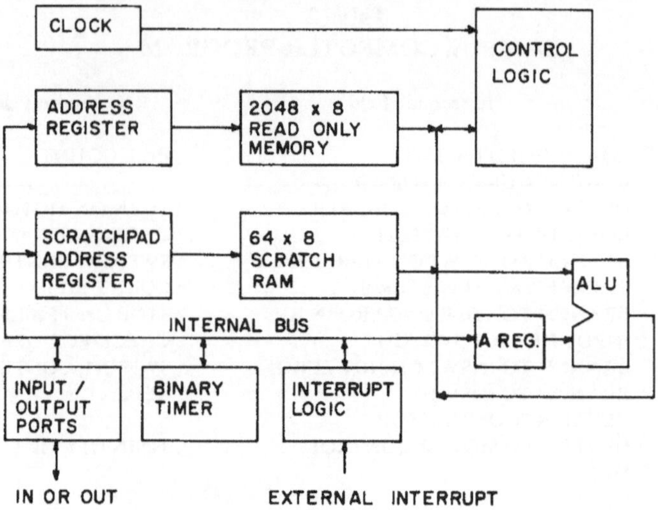

FIGURE 5. Typical single-chip microcomputer.

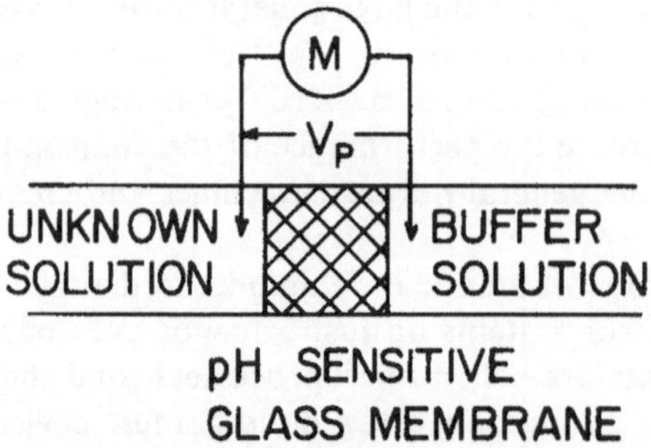

FIGURE 6. The pH electrode.

A. Sequencing and Control

A major application of computers is to control or sequence a process, with or without feedback. In a clinical instrument, a computer can turn on meters, close relays, light lamps, and test control switches. For example, in a blood gas analyzer, the processor can be used to detect the presence of blood at the sample input port and cause the blood to be moved along the measurement path by sequencing motors on pumps and opening and closing valve solenoids. Detectors can be read by the computer to enable it to monitor the flow of the fluid in the system and to provide corrective feedback so that proper flow is maintained. The processor can operate valves that allow wash solutions to clean the sample measurement path and monitor detection circuits that sense a valve malfunction, detect low wash solutions, or other system malfunctions that might affect the quality of the results. Because a processor is programmed, it is possible to change the way that a control sequence operates without changing any of the instrument's hardware. In this way, modifications can be made to add features, improve performance, or correct an error of operation with little cost and effort.

B. Data Input

Most physiological data occur in analog form, i.e., as a continuous function of time. The method used to measure these data usually involves a sensor that provides a voltage or electrical current output which is proportional to the changing physiological parameter. In the pH electrode of Figure 6, a potential difference exists between the two sides of a thin glass membrane when the hydrogen ion concentrations of the two sides are different. If the pH of one of the solutions is held constant (called a buffer), the pH of the other solution can vary, and a variable potential proportional to the unknown solution's pH exists across the membrane.[3]

Digital computers do not operate on analog data, but require numbers that are expressed in digital form. The output of analog sensors cannot be processed directly, but must be converted into a digital format or digitized by a device known as an analog to digital converter (A to D). The instantaneous value of the analog signal is sampled and digitized for entry into the processor. Several types of A to D converters are available, and selection of a type is based on a prior knowledge of the behavior of the signal to be converted. The sensor output as a function of time will determine the conversion rate (conversions per unit time) and the sampling period. The dynamic range of the signal and the required resolution will determine the number of bits required in the converted word. For example, a 12-bit A to D converter can resolve 1 part in 4095, for an accuracy of about 0.025% of full scale.

The most common type of fast analog to digital converter is the successive approximation converter. This method uses a digital to analog converter (DAC) that converts a digital word to an analog voltage, and a voltage comparator that can detect when a measured voltage is greater than the reference voltage generated by the DAC. Using a digital word to control the DAC, the sensor input voltage is compared to the reference voltage starting with the most significant bit. If the input is greater than the reference as determined by the comparator, this bit (MSB) is set to a 1 in the converted word, and the next most significant bit is tested in the same way. If the input is less than the reference generated for any test, that bit is set to a 0. Note that each successive test uses a digital word for the DAC, consisting of that bit and the results of all previous tests, and there are only n tests, where n is the number of bits of conversion resolution.[4]

Another important converter is the dual-slope integrating type. In this method, the input signal is integrated for a fixed period of time with a precision integrator. At the end of this time, the integrator input is switched to a precision reference that causes the integrator output to discharge at a constant rate. When the integrator output is zero, the conversion is complete, and the time required to discharge to zero is proportional to the value of the input signal. When counters are used to generate the input integration period and to measure discharge time, the counters will also contain the converted digital word.[4] This method is much slower than the successive approximation types, is very low cost, tends to reject noise on the inputs, and can be made to have very high resolution.

A third useful conversion technique involves generating a varying frequency signal whose repetition rate is proportional to the input voltage. This technique is useful in applications involving sensor output integration, since the total "counts" over time is the integral.

C. Measurement Techniques

Depending on the measurement being made and the probe type, the useful data might be in the long-term steady-state value of the input signal, in the peak value reached, or in the time period between peaks. A processor can be very helpful by applying consistent rules to extract the data from the input.

A pH electrode has a slow response time to reach a steady-state DC value. Depending on the previous measurement, the electrode can approach this voltage value with either a rising or falling response (see Figure 7). The processor can monitor the slope of the electrode response and detect the end point by looking for zero slope. A time

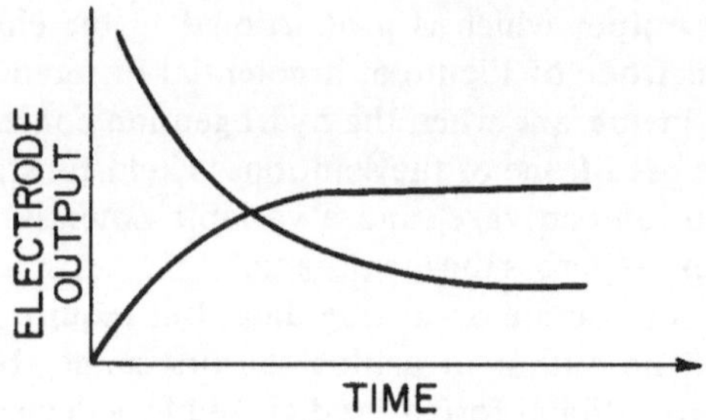

FIGURE 7. pH Electrode response.

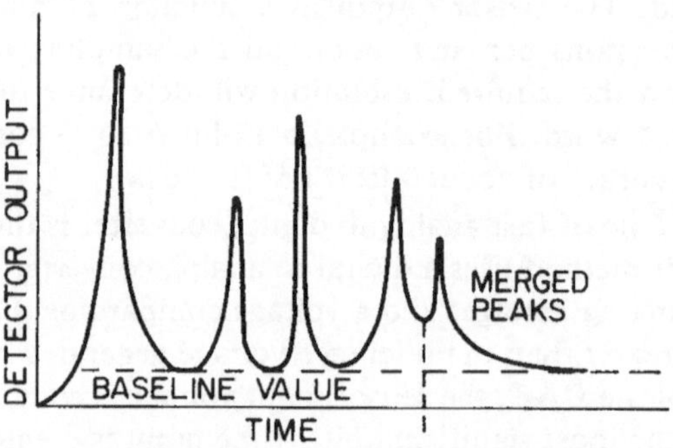

FIGURE 8. Typical gas chromatograph data.

limit could also be used to flag readings that did not reach end point within a maximum time. By averaging successive samples, the output can be filtered to reduce the effects of noise, especially impulse noise.

A peaking response can be monitored for a long-term change in slope, and the maximum or minimum value can be reported.

In a chromatograph, complex chemical mixtures are analyzed and measured by making use of the fact that the different components of the mixture pass through a column of absorbent material at different rates.[5] A detector responds to the presence of the components as they emerge at the end of the column. The detector output is usually recorded on a plotter or strip-chart output device, with the time of emergence being used to identify the chemical component and the area under the peak corresponding to the concentration of the chemical component. Figure 8 shows a typical plot of chromatograph data, with both widely separated peaks and so-called merged peaks. When a processor solution is applied to this measurement problem, sophisticated peak detection methods can be employed, so that small peaks can be detected, while noise interference is reduced. The detector output is digitized and slope is measured by continuously comparing each new sample value to the previous value. If the difference exceeds a certain minimum positive value for several samples, a peak is said to exist, and the accumulation of area continues until a return to baseline is detected. Rapid slope reversals are considered to be noise and are rejected. Noise is reduced by computing a running average value of several sample values. The processor stores the elapsed time since start for each peak detected. Merged peaks are detected when slope changes polarity (+/−) before a previous peak reaches the baseline. The processor may be programmed to assume a valley between peaks and separate areas using this valley. Other separation methods can be programmed as desired. The processor can calculate and report percent of total results and accurate elapsed time since start of measurement

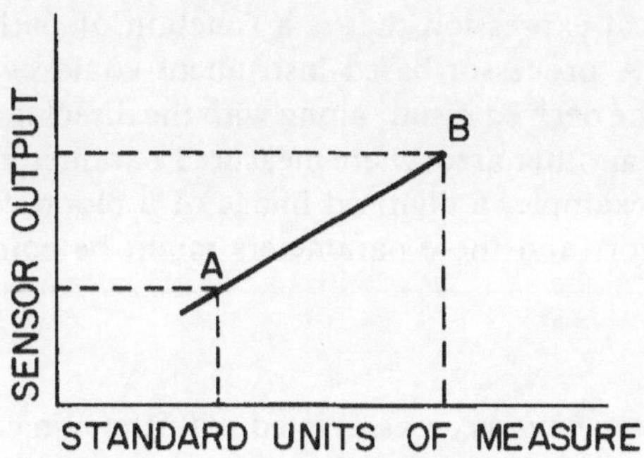

FIGURE 9. Linear sensor calibration.

for each chemical component.[6] By including the ability to enter a known concentration of known substance in the mixture and using a keyboard or other input device, absolute quantity can be determined by normalizing results to the known peak.

D. Calibration

For measurements to be useful, the measuring instrument must be calibrated against a known standard. For many tests, commercially prepared traceable standards are available. Calibration has traditionally been a troublesome area, with results depending on the thoroughness and skill of the operator. Expensive sensors and exotic input circuits are often required to maintain long-term calibration. A processor can simplify the calibration process and store correction factors to permit the use of simple sensors and input circuits. When internal standards are used, calibration can be automatically performed to insure constant quality of results.

The generalized response of a linear sensor is in Figure 9 plotted against units of measure. The equation for a straight line is

$$y = mx + b \quad (3)$$

where m is the slope and b is a constant. The value of m and b can be determined by recording the absolute sensor output in arbitrary units for two known standard samples. The equation is then evaluated by the processor to convert measured values from the sensor into standard units. Nonlinear sensors can be treated as piecewise linear, and those sensors with constant slope can be calibrated with a single sample standard.

E. Derived Results

A common occurance in instrumentation is indirect measurement, where the desired parameter cannot easily be measured, but a mutual effect can. Expressed another way, "what you measure is not what you want." For example, it is easy to measure the time period of a waveform with high accuracy, but signal frequency is the more often desired measurement. Since frequency is the reciprocal of period, a processor can be used to do the division. If a measurement is performed at one set of conditions (temperature, pressure, etc.), a processor can be used to interpolate and provide results for another set of conditions. Often results can be expressed in alternate units such as dB to volts.

Often in medicine a new quantity will be introduced to give a better index of patient condition, and this quantity will be based on a combination of measurable quantities. Base excess is such a quantity that represents the index of metabolic-acid-base imbalance and has a balance value of zero. Metabolic acidosis will produce a negative base excess, and alkalosis will produce a positive base excess.[7] Base excess can be calculated from a complex empirical expression that is a function of patient blood pH, PCO_2, PO_2, and hemoglobin. A processor-based instrument could evaluate the base-excess expression and present the derived result, along with the direct measured results.

Pattern recognition is another area where measured parameters are transformed into more useful form. For example, a digitized image of a blood cell might be measured for size, shape, and color, and these parameters might be combined to detect gross cell abnormalities.

F. Simplification

While digital computers have been associated with the clinical laboratory environment for a number of years, only recently have they become routinely associated with new instruments. The reason is that the cost of processor power in the form of microprocessors and microcomputers has declined to the point where it is feasible to include a computer in most instruments. Previous sections have shown some of the major areas where a processor can be applied to instrument problems. These uses have focused mainly on improved measurement technique, and while they do improve performance, they can be very inefficient in the use of the available computer power.

Whenever a processor is included in a system, there is an overhead cost that is present, regardless of the amount of actual work the processor performs. The more functions provided by the processor, the more cost effective it becomes, as more of the overhead is shared. The cost of expanding the processor is far less than the overhead. It becomes important to not only add features with a processor, but also to study the ways in which all operations of the instrument can be accomplished with the processor. The processor can in most cases replace circuits required by designs without a computer in such areas as data input, operator control, and result output. An example would be replacing expensive thumbwheel switches with a low-cost calculator-like "scanned keyboard" for data entry and control and using the same scanning techniques for output displays. This configuration is not only less costly, but provides a means for the processor to report on instrument condition at no addition in cost.

As an example of simplification, consider an electronic balance that includes a small processor to do the subtraction required by the tare operation of the balance. Such a balance includes a servo-loop to detect balance and an A to D converter to convert the current flowing in the balance coil to a weight. The measured weight is combined with any tare value in the subtraction circuits, and the result is displayed for the operator. Significant simplification can be accomplished by using a digital servo-loop with the processor and using the processor to display the results. Figure 10 shows two versions of the balance and improved processor utilization.

G. Some Limitations

While most instruments can benefit from the application of processor power, some limits to the use of computers do exist. Some measurements require more speed than is now available in processors. Careful attention must also be paid to the range of numbers that can be handled in a program, so that round-off errors do not accumulate. The effects of digital quantizing error due to sampling and conversion must be understood. The digital computer is a very powerful tool, but it must never be considered a cure-all.

V. FUTURE DEVELOPMENTS

The availability of low-cost computers is a recent phenomenon that is already having a profound impact on instrument design. Four main trends are evident in future designs that should all lead to better and often lower-cost health care.

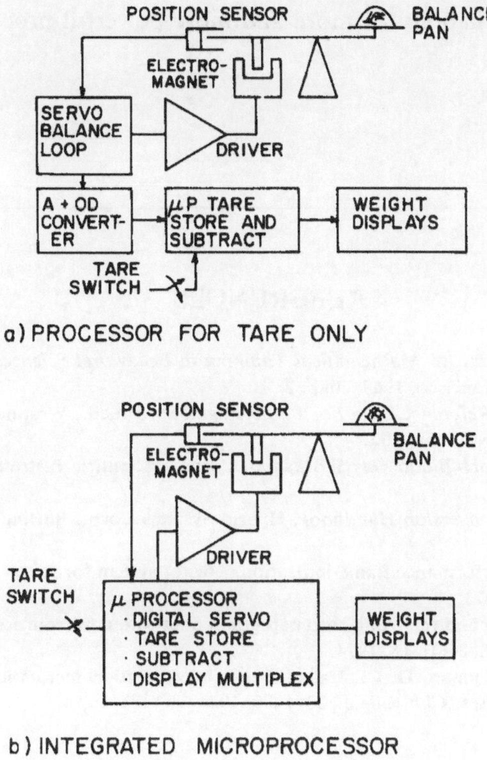

FIGURE 10. Processor based balance.

The first trend is simply the availability of lower-cost instruments which will in turn mean a wider availability of laboratory tests and can result in lower-cost tests. By placing instruments directly in the physician's office, more comprehensive evaluations of patient condition will be encouraged.

The second trend is for smarter instruments that are easy to operate. Not only will these instruments give more consistent results, they will require less operator effort and training. When combined with the first trend, it should be possible to lower patient costs by performing more tests with present staff levels.

The third trend is a new ability to perform complex and sophisticated measurements in a commercial environment with repeatable results. The result is better medicine due to early detection and more complete understanding of patient condition. Many of these techniques will be noninvasive which will lower patient discomfort over present specimen collection methods.

The fourth trend is to design instruments that fit together in a comprehensive measurement system. In larger laboratories, instruments will be connected together, so that an entire battery of tests can be performed on a sample, and the results can be combined and reported together. Automatic sample identification methods will insure accurate results, and more complete testing will provide better health care. Automation should lower costs or at least control them.

VI. CONCLUSION

The digital computer will continue to impact clinical instrumentation by providing new and improved capabilities to measure and by lowering the cost of these instruments. The concepts of the digital computer are not new nor are they complex. What is significant is the availability of more and more powerful processor capabilities at lower and lower costs.

REFERENCES

1. **Ulam, S. M.**, Computers, in *Mathematical Thinking in Behavioral Sciences,* Messick, D. M., Ed., W. H. Freeman, San Francisco, 1968, chap. 2.
2. Digital computers, in *Reference Data For Radio Engineers,* 5th ed., Westman, H. P., Ed., Howard W. Sams, New York, 1972, chap. 32.
3. *Training Manual for pH/Blood Gas Workshop,* Corning Scientific Instruments, Medfield, Mass., 1973.
4. **Bruck, D. B.**, *Data Conversion Handbook,* Hybrid Systems Corp., Burlington, Mass., 1974, chap. 3.
5. **Smith, D. H.**, High performance flame-ionization detector system for gas chromatography, *Hewlett-Packard J.,* 24(7), 2, 1973.
6. **Stefanski, A.**, Deriving and reporting chromatograph data with a microprocessor-controlled integrator, *Hewlett-Packard J.,* 26(4), 18, 1974.
7. **Burnett, R. W. and Noonan, D. C.**, Calculations and correction factors used in determination of blood pH and blood gases, *Clin. Chem.,* 20, 1499, 1974.

THE USE OF HIGH PERFORMANCE LIQUID CHROMATOGRAPHY IN THE CLINICAL LABORATORY

Karen M. Gooding and Fred E. Regnier

TABLE OF CONTENTS

I. Introduction 72

II. Column Selection 73
- A. General 73
- B. Gel Permeation Chromatography 73
- C. Adsorption Chromatography 73
- D. Reversed-Phase Chromatography 73
- E. Ion-Exchange Chromatography 73
- F. Liquid-Liquid Partition Chromatography 74
- G. Ion-Pair Chromatography 74

III. Detection 76
- A. Refractive Index Detectors 76
- B. UV and Visible Photometers 76
- C. Fluorescence Detectors 76
- D. Electrochemical Detection 78
- E. Post Column Reaction 78

IV. Analysis of Normal Constituents of Biological Fluids 79
- A. General Metabolic Profiles 79
- B. Lipids 80
  - 1. General 80
  - 2. Vitamins 80
  - 3. Steroids 80
  - 4. Miscellaneous Lipids 80
- C. Amino Acids 81
- D. Nucleotides 82
- E. Catecholamines 82
- F. Carboxylic Acids 82
- G. Proteins 83
- H. Polyamines 83
- I. Porphyrins 83

V. Drugs and Other Exogenous Substances 84
- A. Poisons, Alkaloids, and Stimulants 84
- B. Sedative-Hypnotic and Anticonvulsant Drugs 85
- C. Antidepressant Drugs and Tranquilizers 86
- D. Theophylline 87
- E. Analgesics and Drugs Used for Gout 88
- F. Drugs for Cardiac Arrhythmias and Anesthetics 89
- G. Antibiotics 89
- H. Miscellaneous Drugs 91

References 94

I. INTRODUCTION

Developing analyses for trace amounts of organic compounds in biological systems is a challenging problem that occupies the time of many scientists and has spawned a number of commercial ventures. Ideally, one would like some simple procedure such as a colorimetric test that could be simply executed in a few minutes. Unfortunately, this is not always possible. With the array of compounds that are encountered in biological systems, simple single-step procedures often lack the specificity and sensitivity for the determination of trace amounts of material. The analysis of tissue, blood, and urine specimens for specific substances often requires some type of preliminary purification.[1] For this reason, chromatographic procedures have played an important role in clinical laboratories.

With the advent of gas chromatography, high-performance chromatographic techniques became available to clinical laboratories. Gas chromatography provided the speed, resolution, and sensitivity that was necessary in many analyses. Unfortunately, gas chromatography has several limitations. First, the direct introduction of biological fluids into gas chromatography columns degrades column performance quickly, and secondly, many polar compounds in body fluids are not volatile. To overcome this problem, it is necessary to go through preliminary sample treatments, such as extraction and solute derivatization. These difficulties in the analysis of polar and high molecular weight materials have contributed to the development of high-performance liquid chromatographic (HPLC) techniques that circumvent the problems of sample pretreatment.

The slower rate of growth in HPLC has been due primarily to the need for more extensive developments in columns and instrumentation than was necessary in gas chromatography. In the case of pumping systems, the construction and refinements of high pressure, nonpulsing, continuous-flow gradient pumping systems that allow the use of a variety of solvents, ranging from buffers to organic solvents, has substantially aided the chromatographer. The expanding array of highly sensitive and selective detectors has also enhanced the use of HPLC.

Some of the most dramatic developments in HPLC have come in the area of column technology. The landmark work of Giddings provided much of the theoretical basis for column development in LC.[2] His theoretical treatment of bandspreading and mass transfer led to the preparation of the pellicular liquid chromatography supports. These packing materials were 37 to 44 μm in diameter, with a rigid, nonporous inner glass core and a thin 1 to 2 μm porous outer layer or pellicle. Solute equilibration between the stationary and mobile phases was quite rapid because the solute only had to diffuse in the thin outer layer of the support. The primary disadvantage of the pellicular supports was that they had limited surface area and therefore low loading capacities (1 to 10 μmol/g).

The discovery by Majors and others of methods to pack microparticulate (5 to 10 μm) supports introduced a new era in liquid chromatography.[3-5] The microparticulate supports had several hundred times greater surface area than the pellicular materials and allowed the construction of 25- to 30-cm columns, with 10,000 or more theoretical plates. Pressure drops on 10-μm particle columns are often less than 1000 psi.

Bonded phase supports have also radically extended the use and convenience of liquid chromatography. Chemical bonding of silane monomers, ranging from the hydrophobic octadecyl silyl group to the very polar hydrophilic γ-aminopropyl silyl group, on the surface of silaceous support materials has provided an array of supports for partitioning solutes from either aqueous or organic mobile phases. The recent introduction of ion-pair partition chromatography is a further refinement of the use of bonded phase supports.

In the past 2 years, column technology and instrumentation have developed to the degree that it is now possible to analyze for trace elements in complex mixtures by HPLC.

II. COLUMN SELECTION

A. General

The selection of a column for HPLC is dependent upon the physical properties of the substance to be analyzed and the medium in which it is found. A variety of methods of separation are available which selectively retain molecules on the basis of charge, polarity, hydrophobicity, size, or solubility.

B. Gel Permeation Chromatography

Perhaps the simplest means of separation is that of size separation or gel permeation chromatogaphy. In this type of chromatography, very large molecules are excluded from the pores of the support, eluting in the column void volume; small molecules are totally included and elute last; and intermediate-sized molecules elute between these limits. Gel permeation chromatography has been used in clinical applications for crude separation of proteins from small molecules. Human serum components have been fractionated by high-performance gel permeation chromatography.[6]

C. Adsorption Chromatography

Adsorption chromatography is the separation of organic soluble compounds by polarity on a surface such as silica or alumina. The composition of the mobile phase and the nature of the adsorbent, including its water content, are critical both for separation and resolution. The silica may be porous with large capacity or pellicular with smaller capacity. Some porous silica materials in general use for HPLC are Zorbax®, Sil-X®, LiChrosorb®, LiChrospher®, Partisil®, μPorasil®, SIL 60®, Vydac TP®, Nucleosil®, and Spherisorb®. Common pellicular materials are Corasil®, Zipax®, Pellosil®, Vydac®, and Pellumina®. Resolution of the methyl substituted purines, theophylline, theobromine, and caffeine by adsorption chromatography is seen in Figure 1.[7]

D. Reversed-Phase Chromatography

Reversed-phase chromatography is the separation of compounds according to their hydrophobic character and their relative solubilities in aqueous and organic solvents. Reversed-phase supports are prepared by binding an aliphatic organosilane to the surface of silica particles. Octadecyl silane (ODS) is the most widely used monomer, but usage of octyl and ethyl bonded phases is increasing. Gradients are run from water or buffer to organic solvents, often methanol or acetonitrile, and compounds elute in order of increasing hydrophobicity. Postscripts such as ODS, C_{18}, RP-18, and RP denote reversed-phase supports. Theophylline, theobromine, and caffeine have also been separated by reversed-phase chromatography as in Figure 2.[8] Theobromine elutes a short time before theophylline in this system.

E. Ion-Exchange Chromatography

In ion-exchange chromatography, separation is based on molecular charge. The support material has a bonded silane with a terminally charged species, such as a sulfonyl or amino group. Elution from an ion-exchange column is usually achieved with an ionic strength and/or pH gradient that selectively displaces charged solutes on the ion exchanger. Some available anion exchangers are found among the Aminex resins, Hamilton resins, and bonded silaceous supports that generally have names with SAX,

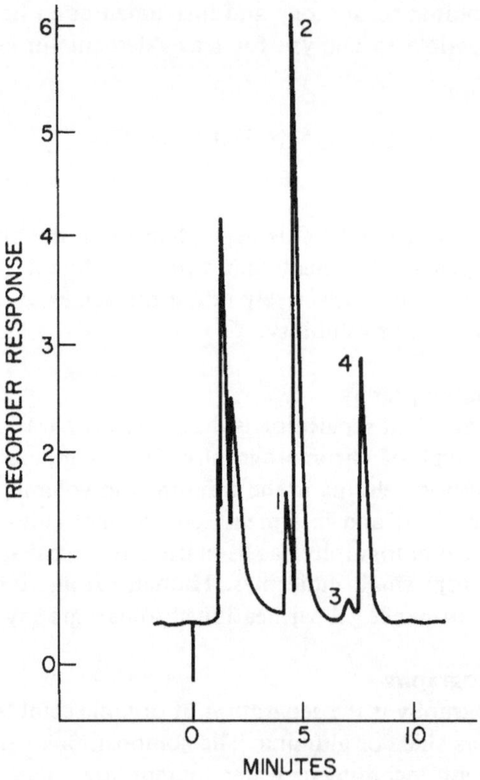

FIGURE 1. Extraction of human serum sample containing caffeine (1), theophylline (2), theobromine (3), and β-hydroxypropyl-theophylline (4). Column, Zorbax-SIL™, 6 to 8 μm; eluent, 6% absolute ethanol, 94% chloroform/heptane/acetic acid (300/200/0.4). (From Evenson, M. A. and Warren, B. L., *Clin. Chem.*, 22, 851, 1976. With permission.)

AAX, or other abbreviations for anion exchange. Cation exchangers are likewise found in both the Aminex and Hamilton series, as well as in bonded silanes with names such as SCX. Figure 3 illustrates the analysis of theophylline, theobromine, and caffeine on Partisil SCX which has a benzene sulfonic acid ion-exchange group.[9]

F. Liquid-Liquid Partition Chromatography

Traditionally, liquid-liquid partition chromatography was practiced by coating the silica support with a viscous liquid and using a mobile phase that was saturated with the coating. Molecules were then distributed between the phases according to solubility. This method was technically difficult, since the stationary phase was continously displaced. The logical solution was to bind the stationary phase to the support material. Whether this is still liquid-liquid partitioning is questionable, but the separations are what would be expected by such a mechanism. Silica which has bonded silane monomers with terminal groups, such as cyano, amino, nitro, diol, or dimethylamino groups, are commercially available.

G. Ion-Pair Chromatography

Ion-pair chromatography is a technique used for the analysis of anionic and cationic compounds. The compounds are paired with reagents which have opposite charges and

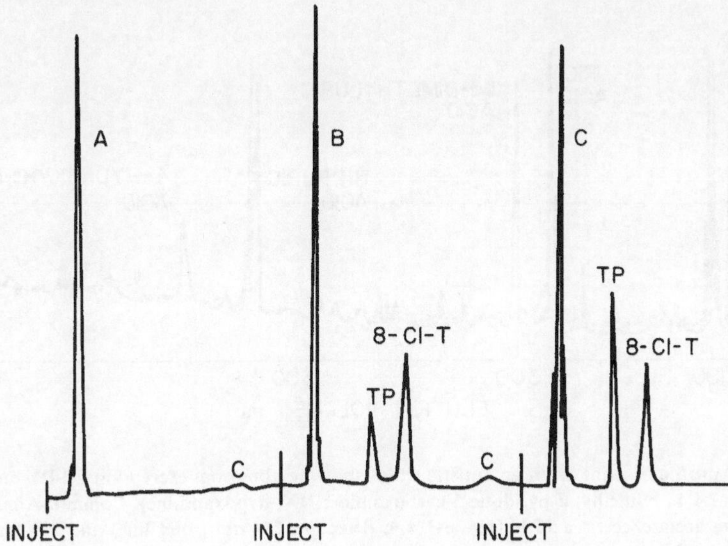

FIGURE 2. Sera extracts. (A) theophylline-free serum without internal standard, (B) patients serum, calculated to contain 7.6 mg/ℓ theophylline, and (C) patient's serum containing 15.1 mg/ℓ (TP, theophylline; 8-Cl-T, internal standard; and C, caffeine.) Column, ODS-Sil-X-I®; eluent, water/acetonitrile/1% acetic acid (480/10/10); detection, 273 nm; temperature, 55°C. (From Adams, R. F., Vandemark, F. L., and Schmidt, G. J., *Clin. Chem.*, 22, 1903, 1976. With permission.)

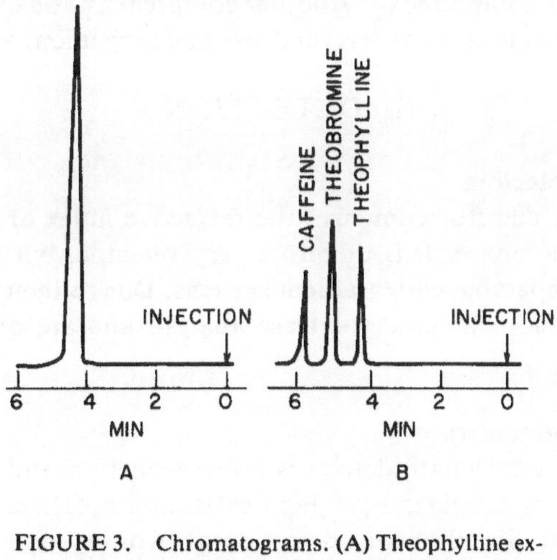

FIGURE 3. Chromatograms. (A) Theophylline extracted from patient's plasma (30.2 μg/mℓ) and (B) theophylline, caffeine, and theobromine extracted from spiked blank plasma (10 μg/mℓ each). Column, Partisil SCX; eluent, 0.66% acetic acid in water; detection, 275 nm; temperature, 50°C. (From Peng, G. W., Smith, V., Peng, A., and Chiou, W. L., *Res. Commun. Chem. Pathol. Pharmacol.*, 15, 341, 1976. PJD Publications, Westbury, N.Y. With permission.)

alkyl chains and then separated. In "straight-phase" ion-pair chromatography, an absorbent surface such as silica is coated with an aqueous stationary phase containing the ion-pairing compound. An organic mobile phase is used for elution. In "reversed-

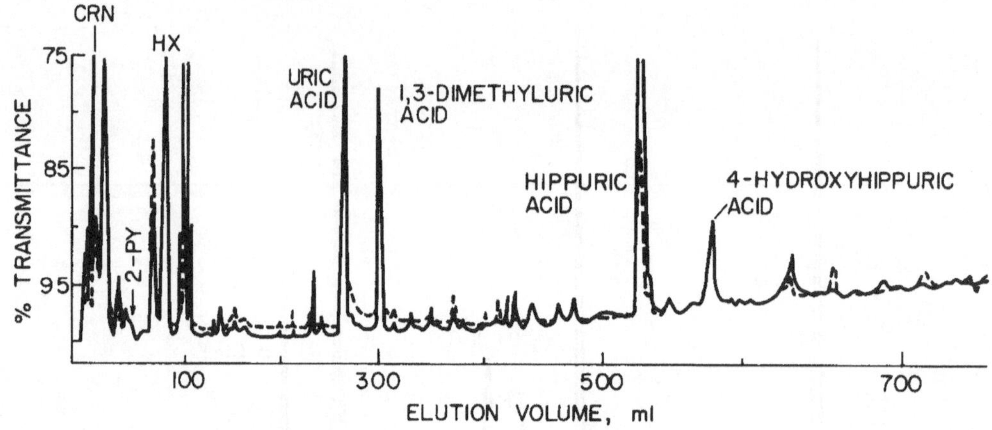

FIGURE 4. UV profile of urine from an autistic child, showing abnormal excretion of 2-PY and purines (CRN, creatinine; 2-PY, N-methyl-2-pyridone-5-carboxamide; HX, hypoxanthine). Column, Aminex A-27; eluent, ammonium acetate/acetic acid buffer, pH 4.4; detection 254 nm (solid line) and 280 nm (broken line). (From Lis, A. W., McLaughlin, D. I., McLaughlin, R. K., Lis, E. W., and Stubbs, E. G., *Clin. Chem.*, 22, 1528, 1976. With permission.)

phase'' ion-pair chromatography, a hydrophobic surface which usually has chain lengths of C_{18}, C_8, or C_2 is coated with an organic stationary phase, such as an intermediate or long-chained alcohol. Solutes are eluted from this column, with an aqueous mobile phase containing the ion-pairing reagent. An explanation of this type of chromatography is included in a study of its use for the determination of acidic drug metabolites and endogenic compounds.[10] Another comprehensive work examined ion-pair chromatography for the analyses of several drugs and biogenic substances.[11]

III. DETECTION

A. Refractive Index Detectors

The refractive index detector compares the refractive index of the column effluent with that of the mobile phase. It is therefore very versatile, but limited, however, in sensitivity and is incompatable with gradient systems. Due to their insensitivity, refractive index detectors cannot be used for trace analysis and are of minimal use in the clinical laboratory.

B. UV and Visible Photometers

The UV and visible wavelength detectors are especially useful in clinical analyses, since many drugs and metabolites have high extinction coefficients at specific wavelengths such as 254 nm. Simultaneous detection at two different wavelengths can increase both specificity and sensitivity, especially in metabolic profiles or other multicomponent mixtures. Figure 4 illustrates dual detection at 254 and 280 nm in a urine analysis.[12] Absorbance ratios at multiple wavelengths have been used to positively identify morphine and methadone in urine.[13] The ratios were obtained by stopping the flow and using a variable wavelength detector to obtain the absorbance readings at 205, 210, 222, and 225 nm.

C. Fluorescence Detectors

Fluorescence detectors are specific and sensitive for compounds which have fluorescent properties. These compounds absorb light at a specific wavelength and then emit light at a longer wavelength. By controlling the wavelength of excitation, the wave-

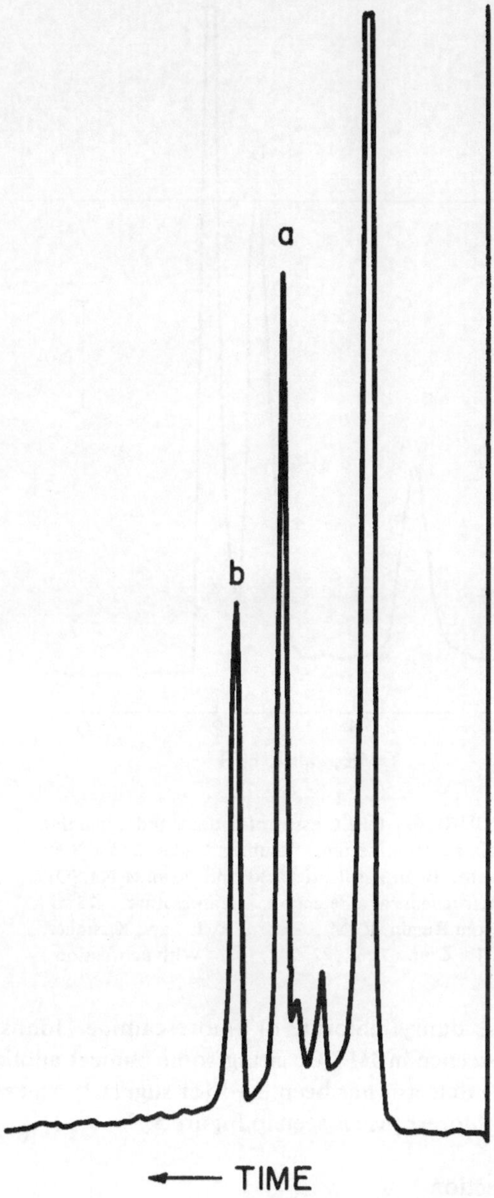

FIGURE 5. HPLC trace of a suspect urine extract, a = LSD and b = iso-LSD. Column, 6 μm Partisil®; eluent, methanol/0.2% ammonium nitrate in water (11/9); detection by fluorescence, λ_{em} = 430 nm, λ_{ex} = 325 nm. (From Christie, J., White, M. W., and Wiles J. M., *J. Chromatogr.*, 120, 496, 1976. With permission.)

length at which emission measurements are made, or both of these, specificity can be attained. The emitted light is monitored at a wavelength different than the excitation frequency and at an angle to the incident light, usually 90°. Therefore, high intensity excitation light sources may be used. Much higher sensitivities may be obtained when one is measuring the intensity of emitted light rather than absorbance.

Fluorescence labels can be introduced into compounds by suitable derivatization

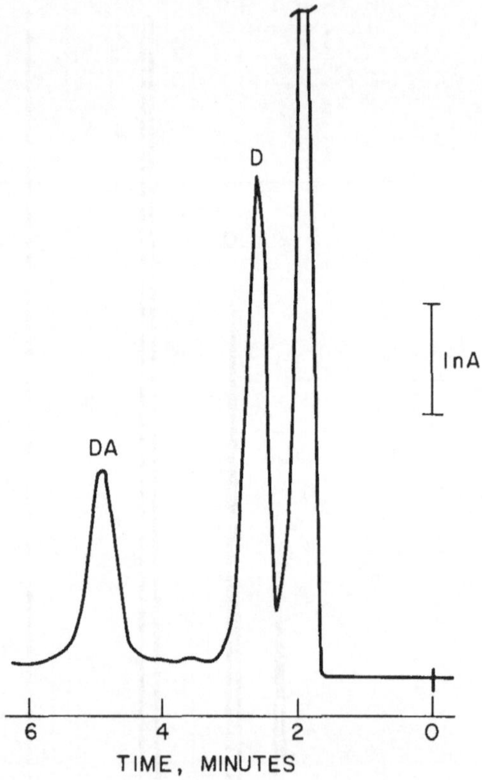

FIGURE 6. HPLC assay of L-dopa and dopamine isolated from serum. Column, Vydac 232 CX®; eluent, 10 m*M* sulfuric acid and 50 m*M* Na_2SO_4; electrochemical detection; temperature, 28°C. (From Riggin, R. M., Alcorn, R. L., and Kissinger, P. T., *Clin. Chem.*, 22, 782, 1976. With permission.)

with compounds such as dansyl chloride or fluorescamine. Johnson et al. have discussed the use of fluorescence in HPLC, citing some clinical applications.[14,15] A free-falling drop fluorescence detector has been used for sugars.[16] An example of the detection of LSD in urine by fluorescence is seen in Figure 5.[17]

D. Electrochemical Detection

Electrochemical detection is a sensitive and selective way to determine electroactive compounds. The detectors can be coulometric, where the conversion efficiency is 100%, or amperometric, where the efficiency is less.[18] One limitation in their use is that the analytical columns must be reversed phase or ion exchange, where it is possible to use mobile phases which are compatible with electrochemistry. Kissinger et al. have used amperometric detection extensively, especially for the analyses of catecholamines[19-21] and aromatic acids.[22-25] An example of the use of amperometric detection is the determination of DOPA and dopamine in Figure 6.[21] Coulometric detection has been used to detect perphenazine and fluphenazine levels in serum.[26]

E. Post Column Reaction

Some compounds have no distinguishing features to enhance their selectivity or to enable their detection at low levels. In some instances, these substances can be derivatized before analysis and the derivatives separated. Alternatively, the effluent can be

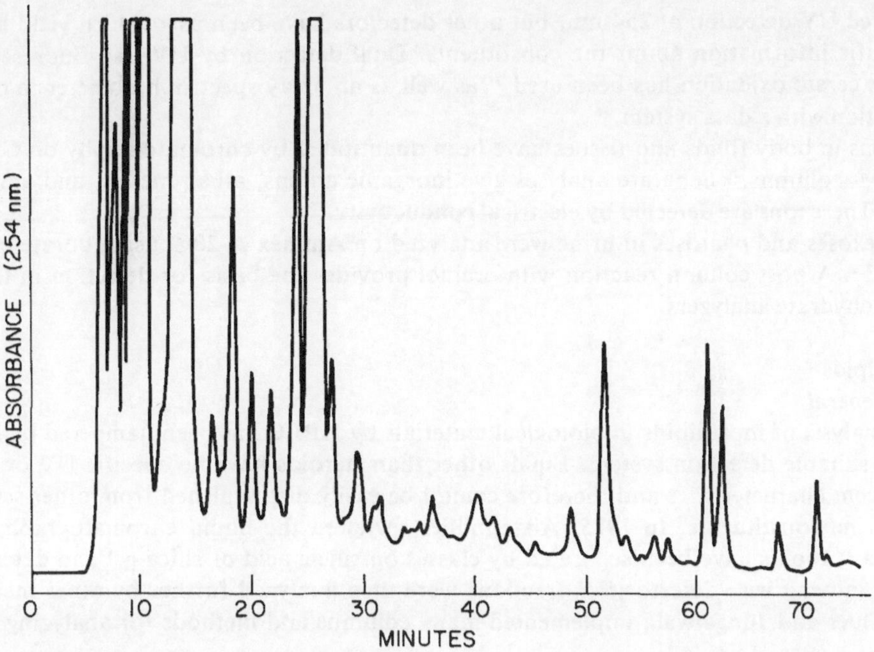

FIGURE 7. Chromatogram of used hemodialysate concentrated tenfold. Column, μ-Bondapak C_{18}™; eluent, sodium acetate gradient to sodium acetate in methanol; detection, 254 nm. (From Senfteber, F. C., Halline, A. G., Veening, H., and Dayton, D. A., *Clin. Chem.*, 22, 1522, 1976. With permission.)

mixed with derivatizing reagent and the derivatized solutes detected. Perhaps the most common example of this technique is the post column reaction with ninhydrin in the amino acid analyzer. In this system, the amino acids and ninhydrin react in a capillary tube. The Technicon Auto Analyzer® makes use of post column reaction in capillaries with air separating segments and minimizing band spreading.[27,28] A third method is that used by Regnier et al. to detect proteins. The column effluent is mixed with substrate solution, and the enzyme reaction products are detected. This reaction is carried out in a column packed with nonporous beads to minimize post column band spreading.[29,30] The isoenzymes of lactate dehydrogenase have been determined in serum by this method.[30] Dansyl chloride, fluorescamine, and fluoropa have been used to convert amines to fluorescent derivatives. Amines have been reacted with ninhydrin after separation on an amino acid analyzer.[31,32] Reaction with orcinol is used to detect carbohydrates,[33] and post column cerate oxidation was used to detect some organic acids.[34]

IV. ANALYSIS OF NORMAL CONSTITUENTS OF BIOLOGICAL FLUIDS

A. General Metabolic Profiles

Variations in metabolic profiles can be indicative of disease and metabolic disorders. A liquid chromatography method which distinguishes a wide spectrum of constituents of urine, serum, or other body fluids will show these variations. Scott developed several methods for metabolic profiling using Aminex A-27, an anion-exchange resin.[35] Others have modified Scott's procedures slightly.[12,34,36] Figure 4 demonstrates this method while contrasting urine profiles of normal and autistic children.[12] These analyses took 40 hr. Recently, the analysis of a hemodialysate by reversed-phase chromatography was performed in only 70 min as in Figure 7.[37] Most of these methods employed UV detection at 254 nm, but other detectors have been used which yield more specific information about the constituents. Dual detection by UV and fluorescence after cerate oxidation has been used,[34] as well as an array spectrophotometer in combination with a data system.[36]

Ions in body fluids and tissues have been quantitated by chromatography on Chromex® columns.[38] Separate analyses give inorganic anions, alkali metals, and ammonia. These ions are detected by electrical conductivity.

Hexoses and pentoses in urine were analyzed on Aminex A-28® using borate buffers.[33,35] A post column reaction with orcinol provides the basis for detection in these carbohydrate analyzers.

B. Lipids

1. General

Analysis of most lipids in biological materials by HPLC has been hampered by lack of a suitable detection system. Lipids other than steroids have no specific UV or fluorescent characteristics and therefore cannot be easily distinguished from other serum or urine constituents. In 1975, Aitzetmüller reviewed the liquid chromatography of lipids.[39] Lipids have been separated by classes on silicic acid or silica gel and detected by a moving wire detector;[40,41] fractions were then analyzed further by other means. McCluer and Jungalwala implemented many columns and methods for analyzing different classes of lipids.[42]

2. Vitamins

Several vitamins have been detected in body fluids by HPLC. These analyses may be of more interest to a nutritionist than a clinical chemist, but some vitamins may be diagnostic. Vitamin A has been analyzed in serum on a MicroPak Si-10® column, with detection at 328 nm.[43] Riboflavin in urine was dually detected by UV at 266 nm and fluorescence after reversed-phase chromatography.[44] Ion-pair chromatography was used for the analysis of nicotinic acid in serum.[45,46] 1,25-Dihydrohyvitamin D was quantitated in human plasma after chromatography on silicic acid, followed by a binding assay on the generated fractions.[47]

3. Steroids

Steroid secretion patterns of human adrenal and testis cells were examined by use of an ODS column.[48] 17-Ketosteroids in plasma and urine were derivatized with dinitrophenylhydrazine and analyzed by liquid-liquid partition chromatography on silica coated with 1.5% ODPN (oxydipropionitrile), as well as by reversed phase.[49] Six estrogens were monitored in human pregnancy urine by chromatography on an amine support.[50] Estrone, estradiol, and estriol have been analyzed by both adsorption and reversed-phase chromatography.[51,52]

Corticosteroid patterns were examined for metabolic disorders by chromatography on an amine support followed by competitive protein binding assays.[53] Prednisolone alone[54] and with other synthetic corticosteroids[55] was analyzed in plasma by Loo and co-workers on LiChrosorb Si60®. Cortisol was quantitatively determined in plasma on Zorbax-Sil®, with simultaneous detection at 239 and 254 nm.[56] Free cortisol has also been quantitated in urine.[57]

Shaw and Elliott tried many chromatography systems to analyze conjugated bile acids; Corasil II® gave the best results with refractive index detection.[58]

4. Miscellaneous Lipids

Neutral glycosphingolipids contained in plasma were benzoylated and chromatographed on Zipax® as seen in Figure 8.[59] Phosphatidyl choline and sphingomyelin

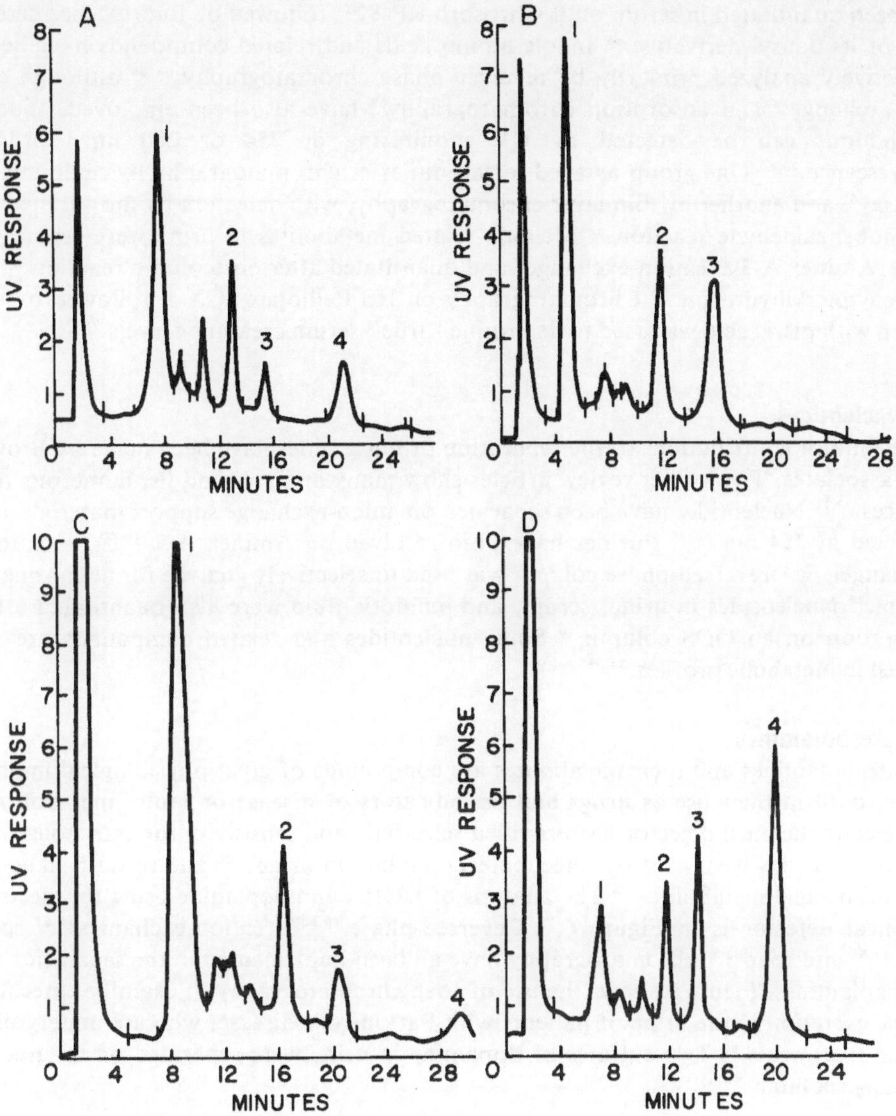

FIGURE 8. HPLC analysis of plasma and erythrocyte glycosphingolipids. A, B, C are results obtained from normal, Fabry's and Gaucher's plasma glycolipid fractions, respectively. Glycolipids from human erythrocytes are labeled D. Peaks 1, 2, 3, and 4 were identified as mono-, di-, tri-, and tetrahexosylceramides, respectively. Column, Zipax™; eluent, methanol in hexane gradient; detection, 254 nm. (From Evans, J. E. and McCluer, R. H., *Biochim. Biophys. Acta*, 270, 565, 1972. With permission.)

have been detected in whole blood and amniotic fluid after separation on MicroPak Si-10®.[60] Two research groups have separated prostaglandins on μ-Porasil® columns with detection by radioactive labeling or analysis of fractions by GC-MS.[61,62]

C. Amino Acids

The amino acid analyzer has been the usual means for quantitating amino acids in biological samples. It employs cation exchange, followed by postcolumn detection with ninhydrin. Möller quantitated amino acids in plasma after ion-exchange chromatography.[63] Ersser tried several polymeric supports and found that 7 to 8% divinylbenzene gave nice resolution of plasma amino acids in 90 to 120 min.[64] ε-Aminocaproic acid has been quantitated in serum on LiChrosorb RP-8®, followed by fluorescence detection of its dansyl derivative.[65] Indole amino acids and related compounds have been collectively analyzed primarily by reversed-phase chromatography,[66-68] although cation-exchange[35] and adsorption chromatography[69] have also been employed. Indole compounds can be detected by UV monitoring at 254 or 280 nm[68] or by fluorescence.[35,66] One group assayed melatonin after chromatography by radioimmunoassay[67] and another by thin layer chromatography, with detection by the 4-dimethylaminobenzaldehyde reaction.[69] Melanin-related metabolites in urine were separated on an Aminex A-5® cation-exchanger and quantitated after postcolumn reaction with diphenylpicrylhydrazine.[70] Chromatography on HS-Pellionex-SCX® followed by reaction with picric acid was used to determine "true" serum creatinine levels.[71]

D. Nucleotides

The major contributions to the separation of nucleotides have been made by Brown and associates. Two of her review articles show many analyses and list numerous references.[72,73] Nucleotides have been separated on anion-exchange support materials and detected at 254 nm.[74-76] Purines have been resolved on Aminex AG-27®, a cation-exchanger.[77] A reversed-phase column was used to selectively analyze for adenosine in serum.[78] Nucleosides in urine, serum, and amniotic fluid were also quantitated after separation on an ODS column.[79] Some nucleotides and related compounds are included in metabolic profiles.[34,35]

E. Catecholamines

Catecholamines and their metabolites are compounds of great physiological importance, both in their use as drugs and as indicators of disease or faulty metabolism. The electrochemical detector has provided selectivity and sensitivity for catecholamine analysis.[23] It has been used to detect catecholamines in urine[19,20] and serum[21] in addition to the acid metabolites.[25] The analysis of DOPA and dopamine using the electrochemical detector is in Figure 6.[21] Reversed-phase,[19,80,81] cation-exchange,[20,21] ion-pair,[11,82] and soap[82,83] chromatography have all been implemented in the separation of catecholamines. Figure 9 shows the use of soap chromatography to examine catecholamine excretion in the urine of patients with Parkinson's disease, who are undergoing varied treatments.[82] Two sulfates of dopamine in urine were separated on an anion-exchange column.[84]

F. Carboxylic Acids

Acids have primarily been resolved on anion-exchange support materials. General metabolic profiles of acids and other material were obtained on Aminex® resins.[12,34,35] Other anion-exchange resins have been implemented to determine homovanillic acid,[25] 3,4 dihydroxyphenylacetic acid in urine,[22] *p*-aminobenzoic acid and metabolites,[24,85,86] and the 2,4 dinitrophenylhydrazine derivatives of pyruvic and α-ketoglutaric acids.[87] The analysis of *p*-aminobenzoic acid (PABA) and its metabolites in serum is in Figure 10.[85] Zipax Permaphase ETH® was used to resolve the naphthalene 2,3-diamine derivative of phenylpyruvic acid.[88] Styrene divinylbenzene was used to distinguish homovanillic from vanilmandelic acid in urine.[89] Figure 11 shows the quantitation of homovanillic acid and vanilmandelic acid in the urine of a patient with neuroblastoma.[89] Liao et al. used an ODS column to determine α-ketoglutaric and pyruvic acid in urine.[90] Molnar and Howath also used reversed-phase chromatography to separate urinary acids; both UV detection at 280 nm and fluorescence were used.[91] Although most of these acids were detected at 254 or 280 nm, homovanillic[25] and *p*-aminobenzoic acids[24] were detected by electrochemical means.

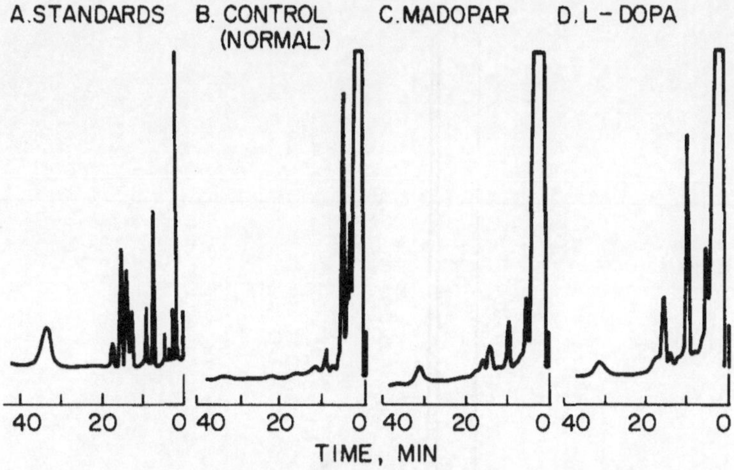

FIGURE 9. Soap chromatography of catecholamines and derivatives in urine. (A) Standards, (B) control normal urine, (C) urine from patient suffering from Parkinson's disease under Madopar treatment, (D) as C, but under L-DOPA treatment. Column, ODS/TMS silica, 5 μm; eluent, water/methanol/sulfuric acid/SLC (72.5/27.5/0.04/0.0005, v/v/v/w); detection, 254 nm. (From Knox, J. H. and Jurand, J., *J. Chromatogr.*, 125, 89, 1976. With permission.)

G. Proteins

Regnier et al. have developed gel permeation and ion-exchange supports which separate proteins and enzymes rapidly without denaturation or loss of activity.[6,29,92,93] Serum proteins and hemoglobins have been separated. Figure 12 shows an analysis of hemoglobins in hemolyzed blood.[29] Lactate dehydrogenase isoenzymes have been analyzed on an anion-exchanger with quantitation by continuous post column reaction with enzyme substrates[30,94] and by assays of individual fractions.[95] Alcohol-soluble proteins, such as albumin, insulin, and histones, have been analyzed on a LiChrosorb Si-60® column.[96]

H. Polyamines

Certain polyamines, such as putrescine, cadaverine, spermidine, and spermine are found in abnormally high concentrations in the biological fluids of cancer patients. Since polyamines cannot be detected with great sensitivity, they must be derivatized. Cation-exchange chromatography with post column derivitization was used to analyze polyamines in urine and whole blood.[31] Reversed-phase chromatography of the dansyl derivatives followed by fluorescence detection was implemented to quantitate polyamines in urine.[14] Gehrke et al. used a small cation-exchange column on an amino acid analyzer as an automated system for quantitation of polyamines in urine, plasma, serum, tissue, and cells.[32]

I. Porphyrins

The patterns of excretion of various porphyrins in urine and feces are indicative of certain disease states and pathological conditions. Porphyrins can be detected at 400 nm, as well as by fluorescence. Profiles of erythropoietic urines were examined using adsorption chromatography and fluorescence detection.[97] Adsorption chromatography of clinical materials with detection at about 400 nm has been used to detect many abnormal porphyrin states.[98-101] Carlson and Dolphin determined porphyrins in fecal material by adsorption chromatography and detection at several wavelengths, 403.5, 300, and 254 nm, to increase selectivity and sensitivity.[102]

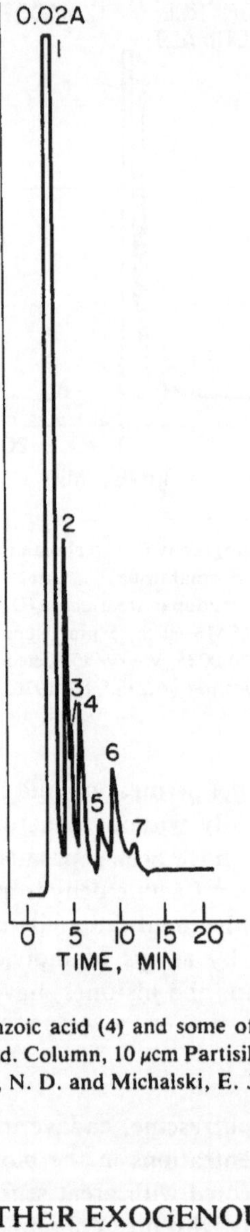

FIGURE 10. Separation of *p*-aminobenzoic acid (4) and some of its metabolites in serum 45 min after oral administration of *p*-aminobenzoic acid. Column, 10 μcm Partisil SAX®; eluent, 0.1 *M* sodium formate, pH 3.5; detection, 254 mn. (From Brown, N. D. and Michalski, E. J., *J. Chromatogr.*, 121, 76, 1976. With permission.)

V. DRUGS AND OTHER EXOGENOUS SUBSTANCES

A. Poisons, Alkaloids, and Stimulants

Paraquot is a herbicide which can cause poisoning and deaths. Pryde and Darby were able to quantitate paraquot in urine after separation on an amine bonded-phase support.[103] Warfarin is an anticoagulant which is used as a drug and also as a rat poison. Warfarin and its metabolites were detected in serum[104] and in plasma[105] after chromatography on an octadecylsilane column with detection at 310 nm.

Many alkaloids, such as morphine, LSD, and heroin, are toxicologically important because they are abused, and drug overdoses are frequent. Wiles et al. analyzed urine for LSD and iso-LSD as in Figure 5.[17] Fluorescence was used for detection after adsorption chromatography. LSD and cannabis in body fluids have been quantitated and their retention characteristics on reversed-phase and cation-exchange columns discussed in detail.[106] Dimerization of morphine with dihydromorphine has been effected by $K_3Fe(CN)_6$ and the resulting three substances quantitated to establish urine levels of morphine.[107] Methadone and morphine levels in urine have been examined using

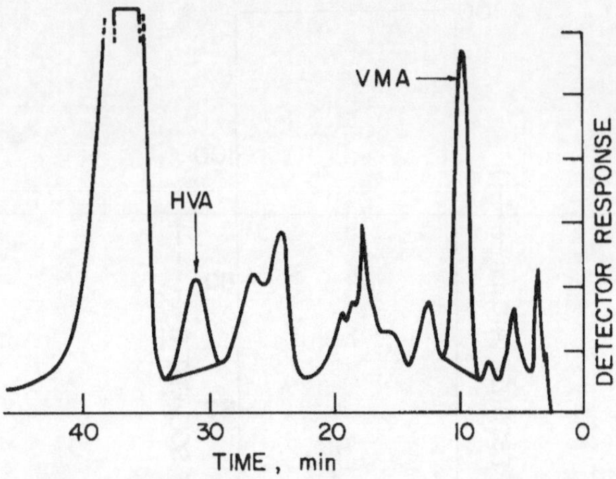

FIGURE 11. Liquid chromatogram of urine from a patient with neuroblastoma. Column, styrene divinylbenzene, 25 μcmc; eluent, 0.05 *M* tartrate/methanol (4/1), pH step gradient; detection, 280 nm. (From Yoshida, A., Yoshioka, M., Tanimura, T., and Tamura, Z., *J. Chromatogr.*, 116, 240, 1976. With permission.)

reversed-phase chromatography with detection by absorbance ratioing.[13] Ergotamine, an alkaloid used for migraine headaches, has been detected in plasma by fluorescence after adsorption chromatography.[108]

Some stimulants have been analyzed by HPLC on silica gel. Nicotine and cotinine in urine were determined at 260 nm.[109] Pemoline in urine has also been quantitated.[110]

B. Sedative-Hypnotic and Anticonvulsant Drugs

The sedative-hypnotic drugs are in wide use as anticonvulsants and sedatives. Although they are effective and necessary drugs, they can cause physical dependence and are liable to abuse. A large dose causes general anesthesia and, ultimately, death.

Tjaden et al. analyzed serum and saliva for most of the barbiturates.[111] Figure 13 shows the analysis of serum on a LiChrosorb Si60® column with bonded dimethyldichlorosilane of a person who took hexobarbital. Phenobarbital, cyclobarbital, heptabarbital, amobarbital, and secobarbital can be easily distinguished with detection at 205 nm in this system. Aprobarbitone, barbitone, and heptabarbitone were analyzed by reversed-phase chromatography of their DANS derivatives with fluorescence detection.[112] Five barbiturates taken in overdose were quantitated in serum by reversed-phase chromatography and detection at 200 nm.[113]

The benzodiazepines are the other major class of sedative-hypnotic drugs that are subject to abuse. Harzer and Barchet used reversed-phase chromatography to separate most of the benzodiazepines and their hydrolysis products in blood and urine.[114] Figure 14 illustrates their technique in the analysis of the blood of a boy who took diazepam. Diazepam and N-desmethyldiazepam were also analyzed in blood by adsorption chromatography and detection at 232 nm.[115] Dopjak et al. determined diazepam, chlordiazepoxide, and their metabolites in plasma and whole blood on a Spherisorb ODS® column, using a 254 nm detector.[116] Nitrazepam and its metabolites in urine were analyzed on a strong anion-exchange column.[117] Flunitrazepam was separated from many other benzodiazepines in plasma by adsorption chromatography with photometric detection at 230 nm.[118]

Carbamazepine and diphenylhydantoin are not sedative-hypnotic drugs, but are also used for anticonvulsant therapy. Carbamazepine was determined in blood and plasma

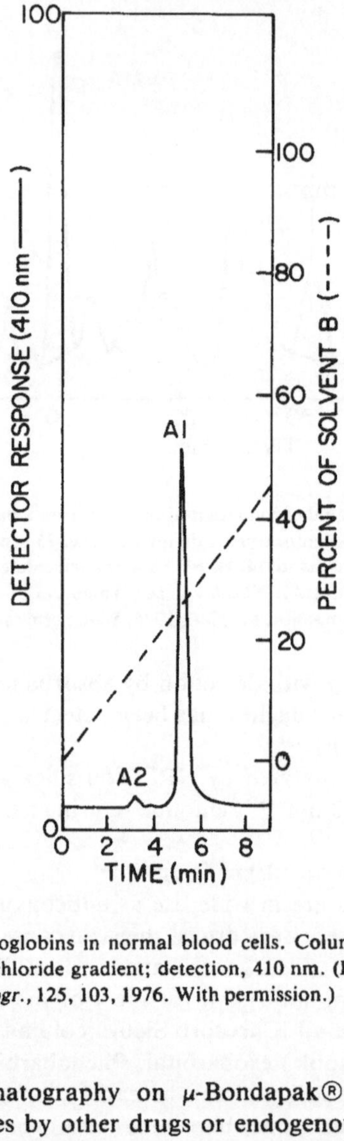

FIGURE 12. Analysis of hemoglobins in normal blood cells. Column, DEAE Glycophase G®; eluent, 0.0125 *M* Tris, pH 8.0, solium chloride gradient; detection, 410 nm. (From Chang, S. H. Gooding, K. M., and Regnier, F. E., *J. Chromatogr.*, 125, 103, 1976. With permission.)

by reversed-phase chromatography on μ-Bondapak® C_{18}[119] and on LiChrosorb RP8®.[120] No interferences by other drugs or endogenous compounds were found in either case. Carbamazepine and its epoxide metabolite were quantitated after chromatography on ODS-Sil-X®[121] and Durapak Carbowax 400-Corasil.®[122] 5-(4-Hydroxyphenyl)-5-phenylhydantoin was assayed in urine by chromatography on μ-Bondapak® C_{18} and detection at 254 nm.[123,124]

Combinations of anticonvulsants have been analyzed. Phenobarbital, diphenylhydantoin, and primidone were quantitated alone on silicic acid,[125] with ethosuximide on ODS-Sil-X-I®,[126] and with ethosuximide and carbamazepine on μ-Bondapak® C_{18}.[127] Finally, all five of these in addition to methsuximide were determined on ODS-Sil-X-I.®[128] Phenobarbital, phenytoin, and carbamazepine and its epoxide were analyzed by adsorption chromatography.[129] These drugs plus primidone and ethosuximide were measured in serum after separation on a reversed-phase column.[130] Ethchlorvynol was determined in rat serum by reversed phase.[131]

C. Antidepressant Drugs and Tranquilizers

Antidepressant drugs and tranquilizers are subject to abuse and often used in suicide attempts. Imipramine, an antidepressant and its metabolite, desmethylimipramine,

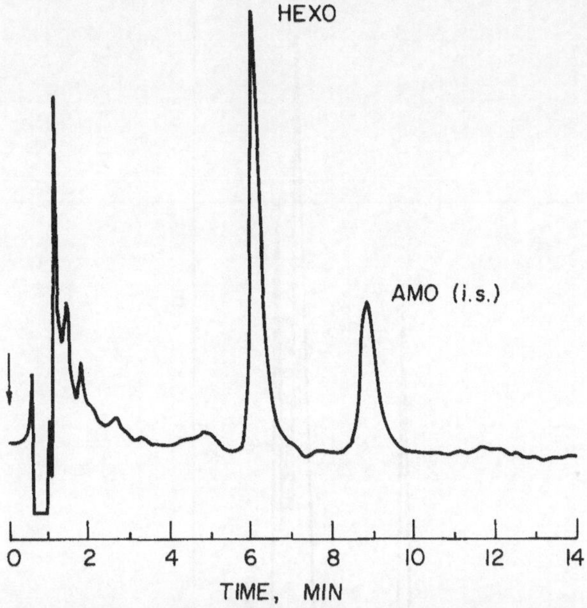

FIGURE 13. Chromatogram of an extract from serum after oral administration of 400 mg of hexobarbital in solid gelatin capsules. Sample taken 6 hr after administration. Column, dimethyl LiChrosorb Si60™; eluent, methanol in water; detection, 205 nm. (From Tjaden, U. R., Kraak, J. C., and Huber, J. F. K., *J. Chromatogr.*, 143, 183, 1977. With permission.)

were analyzed by ion-pair chromatography.[11] Similarly, chlorimipramine and its metabolite were quantitated in plasma by ion-pair chromatography as in Figure 15.[132] Amitryptyline and nortriptyline, two antidepressants, were analyzed in urine and plasma by adsorption chromatography as in Figure 16[133] and by reversed-phase chromatography.[134,135] Amitriptyline and five of its metabolites were determined in serum by chromatography on RP8®.[136]

Chlorpromazine, a tranquilizer, was analyzed in serum by ion-exchange chromatography on ION-X-SC®.[137] Blood levels of perphenazine and fluphenazine were determined by chromatography on methyl-silica with coulometric detection.[26] Thioridazine and its metabolites in blood were analyzed by adsorption chromatography with fluorometric detection after post column reaction.[138]

Watson and Stewart separated several tranquilizers and antidepressants in urine by adsorption chromatography: amitriptyline and its metabolites, imipramine and its metabolites, and chlorpromazine and its metabolites.[139]

D. Theophylline

Theophylline is a vasodilator which is often used to treat asthma attacks. The adverse reactions to this drug can be severe; therefore, it is advantageous to monitor serum levels during administration. Two other xanthines which are found in beverages can interfere with analyses. These are caffeine, a constituent of coffee, tea and, cola drinks, and theobromine, a constituent of cocoa. Theophylline, theobromine, and caffeine in serum or plasma have been analyzed by adsorption,[7,140] cation exchange,[9] and reversed-phase chromatography.[8,141,142] Examples of these were in Figures 1 to 3. Theophylline, dyphylline, theobromine, and phenobarbital in serum have been quantitated using adsorption chromatography,[143] as have theophylline and caffeine in plasma.[144]

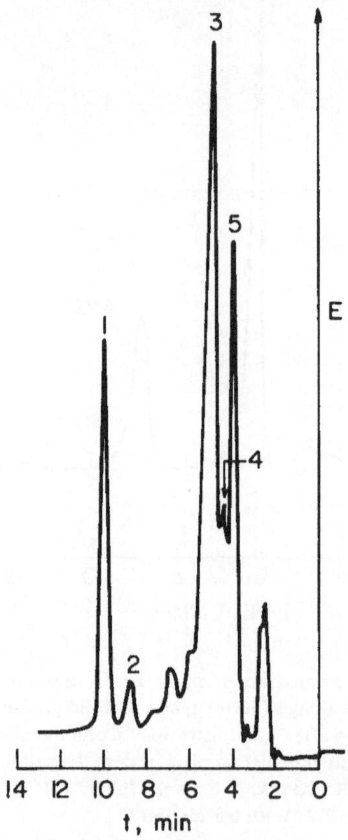

FIGURE 14. Analysis of blood from a boy who took an overdose of diazepam (1). Column, LiChrosorb RP18™; eluent, water to methanol gradient; detection, 254 nm. (From Harzer, K. and Barchet, R., *J. Chromatogr.*, 132, 83, 1977. With permission.)

Theophylline and its metabolites were determined in urine by cation-exchange chromatography on Aminex A-5®.[145] Additionally, theophylline alone has been quantitated in plasma or serum by reversed phase chromatography,[146-150] adsorption chromatography,[151,152] and partition chromatography on Durapak OPN®[152] or MicroPak CN®.[152] It was also determined in saliva after chromatography on μ-Bondapak® C_{18}.[149]

E. Analgesics and Drugs Used for Gout

Salicylic acid has been analyzed in the serum of pediatric patients by reversed-phase chromatography.[154] Salicyluric acid, an aspirin metabolite, was quantitated in urine with fluorometric detection.[106] Acetaminophen, another common analgesic, has been analyzed in serum and plasma by cation-exchange chromatography[155] and in plasma and urine by adsorption chromatography.[156] Acetaminophen and phenacetin were determined in plasma by reversed-phase chromatography.[157-160]

Indomethacin, an analgesic used in extreme circumstances such as gouty arthritis, was analyzed in plasma by reversed-phase chromatography.[161] Sulfinpyrazone, an analgesic and uricosuric agent, has been determined by adsorption chromatography in serum[162] and plasma and urine.[163] Figure 17 illustrates the latter method. Allopurinol and oxipurinol, which reduce uric acid synthesis, were analyzed in plasma and urine by ion-exchange chromatography on Aminex A-27®.[164]

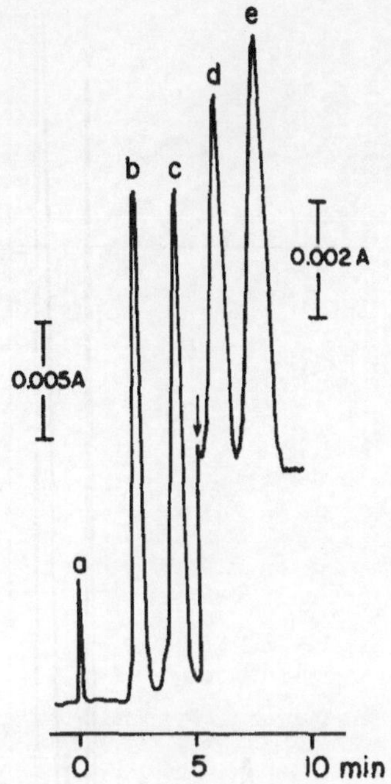

FIGURE 15. Chromatogram of a plasma sample from a patient receiving chlorimipramine hydrochloride (a, sample injection; b, solvent front; c, desmethylchlorimipramine; d, chlorimipramine; and e, trimipramine). Column, DiaChrom® coated with 25% (w/v) of a solution of 0.1 *M* HCl and 0.1 *M* tetraethylammonium chloride; 37 to 44 μm; eluent, hexane-isobutanol (9/1); detection, 255 nm. (From Mellström, B. and Eksborg, S., *J. Chromatogr.*, 116, 475, 1976. With permission.)

F. Drugs for Cardiac Arrhythmias and Anesthetics

Antiarrhythmic drugs are local anesthetics which must be used with caution; therefore, a rapid means of monitoring their levels in body fluids is necessary. Quinidine has been analyzed by ion-pair chromatography.[11] Procainamide is another local anesthetic used to treat cardiac arrhythmias. In addition to monitoring levels of this drug, its acetylated derivative can be determined and is indicative of the patient's acetylation rate. Procainamide and N-acetylprocainamide in serum and plasma have been analyzed on Zorbax-SIL[165]® and on μ-Bondapak® C_{18}.[166-168] An example of the latter method is in Figure 18.[165] Lidocaine has also been separated from procainamide by reversed-phase chromatography.[169] Disopyramide and its mono-N-dealkylated metabolite were analyzed in plasma and urine by means of ion-pair chromatography with detection at 254 nm.[170] Acebutolol and its major metabolite in blood, plasma, and urine were analyzed by soap chromatography on MCH-10®.[171]

Benzoylecgonine, a polar metabolite of cocaine, was isolated from urine by reversed-phase chromatography, followed by reanalysis by gas chromatography-mass spectrometry.[172] Ketamine, a general anesthetic, and some of its metabolites were analyzed in rat urine by reversed-phase chromatography.[173]

G. Antibiotics

The clinical significance of the determination of antibiotics in biological fluids may be questionable, since such assays are usually used in pharmacological studies. However, for the sake of continuity, they are included.

A broad spectrum antibiotic, β-Cetotetrine, was analyzed in plasma and urine using

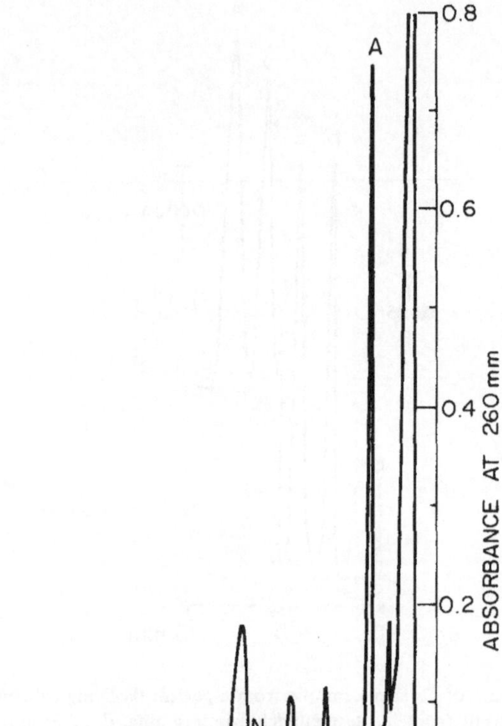

FIGURE 16. Trace obtained from the urine of a patient 6 hr after taking an overdose of amitriptyline (A, amitriptyline; N, nortriptyline; and DMI, desmethylimipramine). The peak immediately behind nortriptyline is 10-hydroxyamitriptyline, and the peak following amitriptyline is believed to be desmethylnortriptyline.[195] Column, 5 μcm silica; eluent, dichloromethane-isopropanol-concentrated ammonia (100/2/0.25); detection, 240 nm. (From Watson, I. D. and Stewart, M. J., *J. Chromatogr.*, 132, 155, 1977. With permission.)

reversed-phase chromatography with Bondapak Phenyl/Corasil® followed by thin-layer electrochemistry for detection.[174] Several studies have examined cephalosporins in biological fluids. Cephalothin and its deacetylated metabolite in serum and urine were analyzed by ion-pair extraction and ion-exchange chromatography on AS-Pellionex-SAX®.[175] Cefoxitin, cephalothin, and their deacetylated metabolites were separated from other urine constituents by ion-exchange on Zipax SAX®.[176] Cephalosporin HR 580® in plasma and urine was analyzed by reversed-phase chromatography.[177]

Daunomycin and daunomycinol, which are used in acute leukemias, were determined in plasma by chromatography on Zorbax-SIL®.[178] Ftorafur, an anticancer drug used for breast and GI malignancies, was analyzed with its metabolites by reversed-phase chromatography.[179]

Nalidixic acid, which is used for urinary tract infections, and hydroxynalidixic acid were analyzed in plasma and urine by anion-exchange chromatography.[180] Two antileprotic and antimalarial sulfones, diaminodiphenysulfone (DDS) and acedapsone,

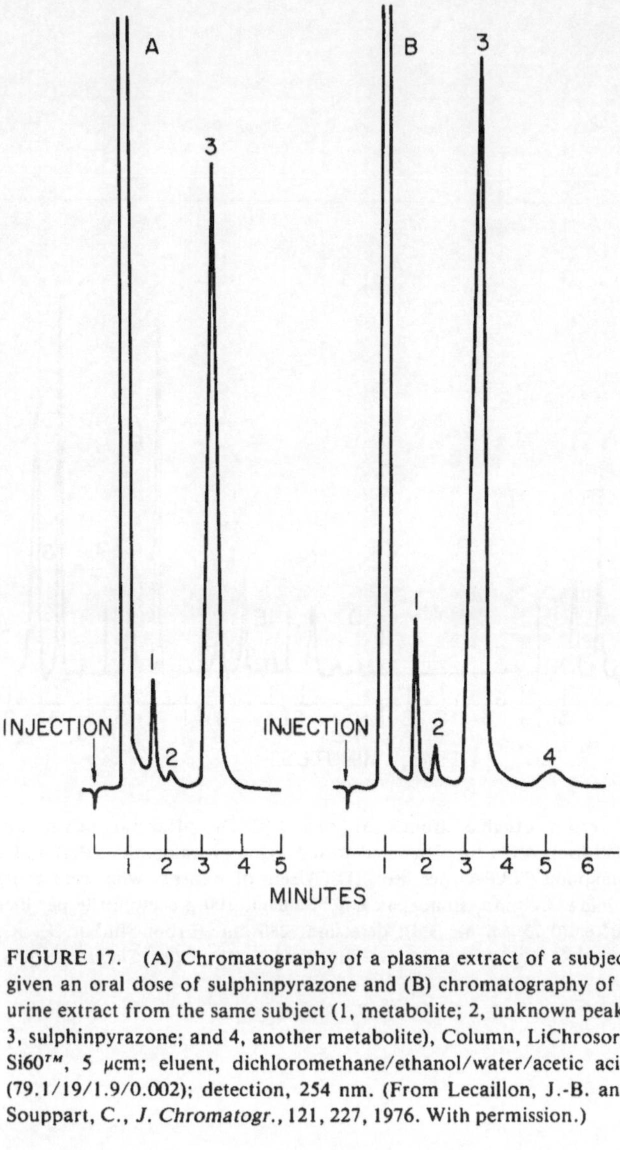

FIGURE 17. (A) Chromatography of a plasma extract of a subject given an oral dose of sulphinpyrazone and (B) chromatography of a urine extract from the same subject (1, metabolite; 2, unknown peak; 3, sulphinpyrazone; and 4, another metabolite), Column, LiChrosorb Si60™, 5 μcm; eluent, dichloromethane/ethanol/water/acetic acid (79.1/19/1.9/0.002); detection, 254 nm. (From Lecaillon, J.-B. and Souppart, C., *J. Chromatogr.*, 121, 227, 1976. With permission.)

plus the plasma metabolite of DDS, were analyzed on silica gel with fluorescence detection.[181] An antifungal agent, griseofulvin, was determined in urine, as its metabolite by adsorption chromatography on Pellosil.®.[182]

H. Miscellaneous Drugs

Tolbutamide, a hypoglycemic drug, has been analyzed in serum using liquid-liquid partion chromatography on ETH Permaphase®.[183] Tolbutamide and chlorpropamide in plasma were determined by reversed-phase chromatography.[184] Figure 19 shows the analysis of the plasma of a patient taking tolbutamide using this technique. Urine samples containing metformin, another hypoglycemic agent, were derivatized with *p*-nitrobenzoyl chloride and then analyzed on a Bondapak Phenyl/Corasil® column.[185] Mefruside, a diuretic and antidiabetic drug, and its metabolites were determined in urine by adsorption chromatography.[186]

Clofibrinic acid, a hypolipidaemic drug, was determined in plasma, urine, and saliva by analysis of MicroPak CH-10®.[187] *p*-Chlorophenoxyisobutyric acid, the product of

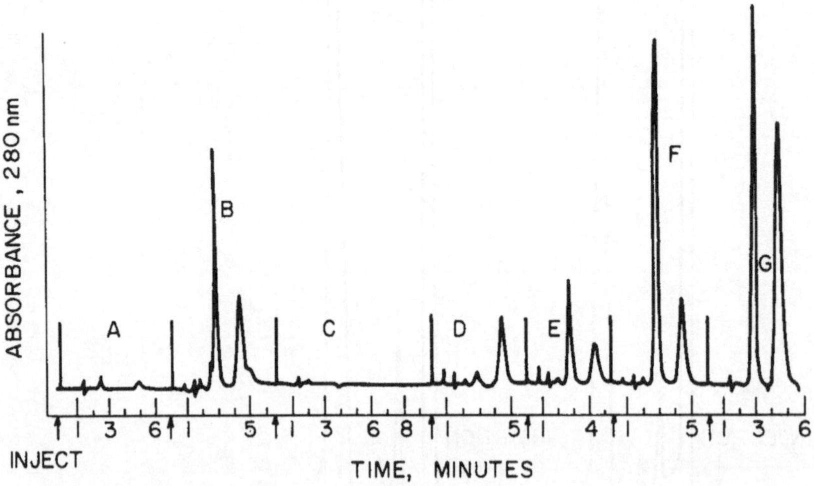

FIGURE 18. HPLC records obtained from sera. (A) and (C) Drug-free patient's serum, (B) drug-free serum ("Monitrol®) supplemented with 5 μg of procainamide (PA) and 5 mg of N-acetylprocainamide (NAPA) per liter, (D-G) sera of patients who were being treated with procainamide. Column, Bondapak C_{18}™; eluent, 100 g acetonitrile per liter of sodium acetate buffer (0.75 *M*, pH 3.4); detection, 280 nm. (From Shukur, L. R., Powers, J. L., Marques, R. A., Winter, M. E., and Sandee, W., *Clin. Chem.*, 23, 636, 1977. With permission.)

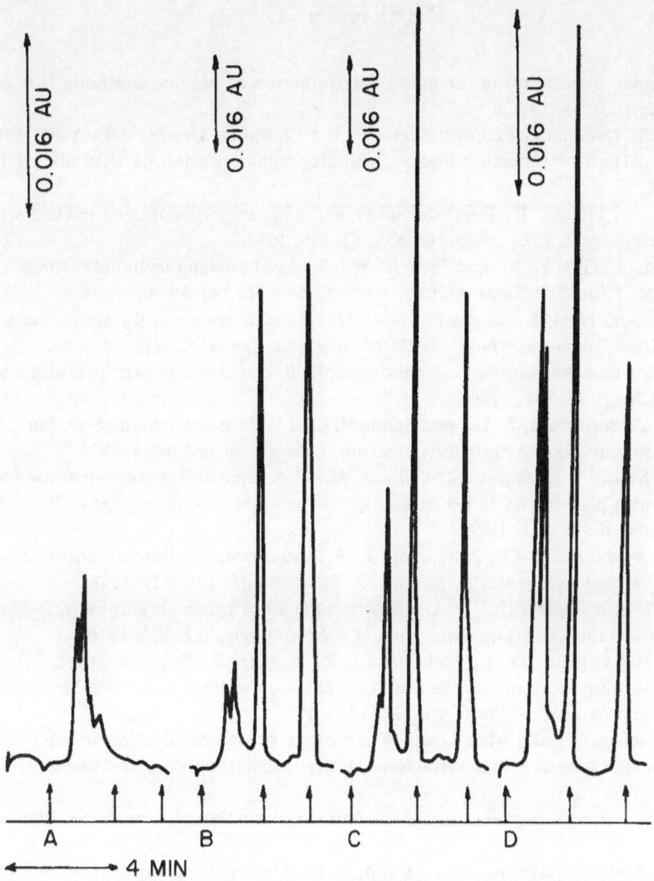

FIGURE 19. Chromatograms of chlorpropamide and tolbutamide extracted from plasma. (A) Blank plasma, (B) blank plasma spiked with 20 μg chlorpropamide and 100 μg tolbutamide per milliliter (C) plasma from an individual medicated with chlorpropamide (estimated 31.1 μg/mℓ, with tolbutamide as an internal standard (100 μg/mℓ), and (D) plasma from an individual medicated with tolbutamide (estimated 56.7 μg/mℓ), with chlorpropamide as an internal standard (20 μg/mℓ). Column, μ-Bondapak C_{18}™; eluent, 17% acetonitrile in 0.05 *M* aqueous ammonium formate; detection, 254 nm. (From Sved, S., McGilveray, I. J., and Beaudoin, N., *J. Pharm. Sci.*, 65, 1356, 1976. With permission.)

the hydrolysis of clofibrate, was quantitated in plasma after reversed-phase chromatography.[188]

Hydrochlorothiazide, a diuretic, has been determined by reversed-phase chromatography in serum[190-192] as well as urine.[191,192] Methimazole, an antithyroid agent, has been determined in the plasma of thyrotoxic patients by chromatography on Spherisorb-Al® or Spherisorb-Si®.[193] Diftalone, an anti-inflammatory drug, and its metabolites in plasma and synovial fluid have been determined by reversed-phase chromatography.[194]

REFERENCES

1. **Reid, E.,** Sample preparation in the micro-determination of organic compounds in plasma or urine, *Analyst London,* 101, 1, 1976.
2. **Giddings, J. C.,** *Dynamics of Chromatography, Part I,* Marcel Dekker, New York, 1965.
3. **Majors, R. E.,** High performance liquid chromatography on small particle silica gel, *Anal. Chem.,* 44, 1722, 1972.
4. **Asshauer, J. and Halasz, I.,** Reproducibility and efficiency of columns packed wtih 10μ silica in liquid chromatography, *J. Chromatogr. Sci.,* 12, 139, 1974.
5. **Cassidy, R. M., LeGay, D. S., and Frei, R. W.,** Study of packing techniques for small-particle silica gels in high-speed liquid chromatography, *Anal. Chem.,* 46, 340, 1974.
6. **Regnier, F. E. and Noel, R.,** Glycerolpropylsilane bonded phases in the steric exclusion chromatography of biological macromolecules, *J. Chromatogr. Sci.,* 14, 316, 1976.
7. **Evenson, M. A. and Warren, B. L.,** Serum theophylline analysis by high-pressure liquid chromatography, *Clin. Chem.,* 22, 851, 1976.
8. **Adams, R. F., Vandemark, F. L., and Schmidt, G. J.,** More sensitive high-pressure liquid-chromatographic determination of theophylline in serum, *Clin. Chem.,* 22, 1903, 1976.
9. **Peng, G. W., Smith, V., Peng, A., and Chiou, W. L.,** A rapid and sensitive method for determination of theophylline in plasma and saliva by high pressure liquid chromatography, *Res. Commun. Chem. Pathol. Pharmacol.,* 15, 341, 1976.
10. **Fransson, B., Wahlund, K.-G., Johansson, I. M., and Schill, G.,** Ion-pair chromatography of acidic drug metabolites and endogenic compounds, *J. Chromatogr.,* 125, 327, 1976.
11. **Persson, B.-A. and Lagerström, P.-O.,** Ion-pair partition chromatography in the analysis of drugs and biogenic substances in plasma and urine, *J. Chromatogr.,* 122, 305, 1976.
12. **Lis, A. W., McLaughlin, D. I., McLaughlin, R. K., Lis, E. W., and Stubbs, E. G.,** Profiles of ultraviolet-absorbing components of urine from autistic children, as obtained by high-resolution ion-exchange chromatography, *Clin. Chem.,* 22, 1528, 1976.
13. **Yost, R., Stoveken, J., and Mac Lean, W.,** Positive peak identification in liquid chromatography using absorbance ratioing with a variable-wavelength spectrophotometric detector, *J. Chromatogr.,* 134, 73, 1977.
14. **Johnson, E. L.,** *Fluorescence and Liquid Chromatography,* Varian Aerograph, Walnut Creek, Calif., 1977.
15. **Johnson, E., Abu-Shumays, A., and Abbott, S. R.,** Use of fluorescence detection in high-performance liquid chromatography, *J. Chromatogr.,* 134, 107, 1977.
16. **Martin, F., Maine, J., Sweeley, C., and Holland, J. F.,** The free-falling drop detector — a novel fluorescence detector for high-performance liquid chromatography, *Clin. Chem.,* 22, 1434, 1976.
17. **Christie, J., White, M. W., and Wiles, J. M.** A chromatographic method for the detection of LSD in biological liquids, *J. Chromatogr.,* 120, 496, 1976.
18. **Kissinger, P. T.,** Amperometric and coulometric detectors for high-performance liquid chromatography, *Anal. Chem.,* 49, 247, 1977.
19. **Shoup, R. E. and Kissinger, P. T.,** Determination of urinary normetanephrine, metanephrine, and 3-methoxytyramine by liquid chromatography, with amperometric detection, *Clin. Chem.,* 23, 1268, 1977.
20. **Kissinger, P. T., Riggin, R. M., Alcorn, R. L., and Rau, L. D.,** Estimation of catecholamines in urine by high performance liquid chromatography with electrochemical detection, *Biochem. Med.,* 13, 299, 1975.
21. **Riggin, R. M., Alcorn, R. L., and Kissinger, P. T.,** Liquid chromatographic method for monitoring therapeutic concentrations of L-dopa and dopamine in serum, *Clin. Chem.,* 22, 782, 1976.
22. **Felice, L. J., Bruntlett, C. S., and Kissinger, P. T.,** Liquid chromatography assay for 3,4-dihydroxyphenylacetic acid in urine, *J. Chromatogr.,* 143, 407, 1977.
23. **Kissinger, P. T., Bruntlett, C. S., Davis, G. C., Felice, L. J., Riggin, R. M., and Shoup, R. E.,** Recent developments in the clinical assessment of aromatic metabolism by high performance reverse-phase chromatography with amperometric detection, *Clin. Chem.,* 23, 1449, 1977.
24. **Shoup, R. E. and Kissinger, P. T.,** A simple liquid chromatography procedure for *p*-aminohippuric acid in blood serum and urine, *Biochem. Med.,* 14, 317, 1975.
25. **Felice, L. J. and Kissinger, P. T.,** Determination of homovanillic acid in urine by liquid chromatography with electrochemical detection, *Anal. Chem.,* 48, 794, 1976.
26. **Tjaden, U. R., Lankelma, J., Poppe, H., and Muusze, R. G.,** Anodic coulometric detection with a glassy carbon electrode in combination with reversed-phase high-performance liquid chromatography. Determination of blood levels of perphenazine and fluphenazine, *J. Chromatogr.,* 125, 275, 1976.

27. **Snyder, L. R. and Adler, H. J.,** Dispersion in segmented flow through glass tubing in continuous-flow analysis: the ideal model, *Anal. Chem.*, 48, 1017, 1976.
28. **Snyder, L. R. and Adler, H. J.,** Dispersion in segmented flow through glass tubing in continuous-flow analysis: the nonideal model, *Anal. Chem.*, 48, 1022, 1976.
29. **Chang, S. H., Gooding, K. M., and Regnier, F. E.,** High-performance liquid chromatography of proteins, *J. Chromatogr.*, 125, 103, 1976.
30. **Schlabach, T. D., Chang, S. H., Gooding, K. M., and Regnier, F. E.,** A continuous-flow enzyme detector for liquid chromatography, *J. Chromatogr.*, 134, 91, 1977.
31. **Adler, H., Margoshes, M., Snyder, L. R., and Spitzer, C.,** Rapid chromatographic method to determine polyamines in urine and whole blood, *J. Chromatogr.*, 143, 125, 1977.
32. **Gehrke, C. W., Kuo, K. C., and Ellis, R. L.,** Polyamines — an improved automated ion-exchange method, *J. Chromatogr.*, 143, 345, 1977.
33. **Morrison, W. H., Lou, M. F., and Hamilton, P. B.,** The determination of hexoses and pentoses by anion-exchange chromatography: a method of high sensitivity, *Anal. Biochem.*, 71, 415, 1976.
34. **Katz, S., Pitt, W. W., and Mrochek, J. E.,** Comparative serum and urine analysis by dual-detector anion-exchange chromatography, *J. Chromatogr.*, 104, 303, 1975.
35. **Scott, C. D.,** High-pressure ion exchange chromatography as applied to the separation of complex biochemical mixtures, *Sep. Purif. Methods*, 3, 263, 1974.
36. **Dessey, R. E., Reynolds, W. D., Nunn, W. G., Titus, C. A., and Moler, G. F.,** A minicomputer-automated array spectrometer for liquid chromatographic detection of metabolites, *Clin. Chem.*, 22, 1472, 1976.
37. **Senfteber, F. C., Halline, A. G., Veening, H., and Dayton, D. A.,** Reversed-phase liquid chromatographic analysis of hemodialysate from uremic patients, *Clin. Chem.*, 22, 1522, 1976.
38. **Anderson, C.,** Ion chromatography: a new technique for clinical chemistry, *Clin. Chem.*, 22, 1424, 1976.
39. **Aitzetmüller, K.,** The liquid chromatography of lipids. A critical review, *J. Chromatogr.*, 113, 231, 1975.
40. **Worth, H. G. J.,** Some early experience with a moving wire liquid chromatograph, in *High Pressure Liquid Chromatography in Clinical Chemistry*, Dixon, P. F., Gray, C. H., Lim, C. K. and Stoll, M. S., Eds., Academic Press, London, 1976, 11.
41. **Carter, T., Magnani, H. N., and Watts, R.,** Analysis of serum lipids and lipoproteins by high speed liquid chromatography, in *High Pressure Liquid Chromatography in Clinical Chemistry*, Dixon, P. F., Gray, C. H., Lim, C. K., and Stoll, M. S., Eds., Academic Press, London, 1976, 33.
42. **McCluer, R. H. and Jungalwala, F. B.,** High-performance liquid chromatographic analysis of glycosphingolipids and phospholipids, *Adv. Exp. Med. Biol.*, 68, 533, 1976.
43. **DeRuyter, M. G. M. and DeLeenheer, A. P.,** Determination of serum retinol (Vitamin A) by high-speed liquid chromatography, *Clin. Chem.*, 22, 1593, 1976.
44. **Rhys Williams, A. T. and Slavin, W.,** Determination of riboflavin in milk and riboflavin clearance into urine using HPLC with fluorescence detection, *Chromatogr. Newsl.*, 5, 9, 1977.
45. **Wahlund, K.-G.,** Reversed-phase ion-pair partition chromatography of carboxylates and sulphonates, *J. Chromatogr.*, 115, 411, 1975.
46. **Wahlund, K.-G. and Lund, U.,** Reversed-phase ion-pair partition chromatography of phenylacetic, mandelic and pyridinecarboxylic acid derivatives, *J. Chromatogr.*, 122, 269, 1976.
47. **Eisman, J. A., Hamstra, A. J., Kream, B. E., and DeLuca, H. F.,** A sensitive, precise and convenient method for determination of 1,25 dihydroxyvitamin D in human plasma, *Arch. Biochem. Biophys.*, 176, 235, 1976.
48. **O'Hare, M. J., Nice, E. C., Magee-Brown, R., and Bullman, H.,** High-pressure liquid chromatography of steroids secreted by human adrenal and testis cells in monolayer culture, *J. Chromatogr.*, 125, 357, 1976.
49. **Fitzpatrick, F. A., Siggia, S., and Dingman, J.,** High speed liquid chromatography of derivatized urinary 17-keto steroids, *Anal. Chem.*, 44, 2211, 1972.
50. **Fantl, V., Lim, C. K., and Gray, C. H.,** Separation of oestrogens and determination of oestriol in human pregnancy urine, in *High Pressure Liquid Chromatography in Clinical Chemistry*, Dixon, P. F., Gray, C. H., Lim, C. K., and Stoll, M. S., Eds., Academic Press, London, 1976, 51.
51. Oestrogenic steroids in pregnancy urine, in *Liquid Chromatography Applications. Biomedical Analysis*, Pye Unicam, Cambridge, 1976.
52. **Dolphin, R. J. and Pergande, P. J.,** Improved method for the analysis of estrogenic steroids in pregnancy urine by high-performance liquid chromatography, *J. Chromatogr.*, 143, 267, 1977.
53. **Butler, J., Fantl, V., and Lim, C. K.,** Separation of corticosteroids for measurement by competitive protein binding, in *High Pressure Liquid Chromatography in Clinical Chemistry*, Dixon, P. F., Gray, C. H., Lim, C. K., and Stoll, M. S., Eds., Academic Press, London, 1976, 59.

54. **Loo, J. C. K., Butterfield, A. G., Moffatt, J., and Jordan, N.**, Analysis of prednisolone in plasma by high-performance liquid chromatography, *J. Chromatogr.*, 143, 275, 1977.
55. **Loo, J. C. K. and Jordan, N.**, High-performance liquid chromatographic analysis for synthetic corticosteroids in plasma, *J. Chromatogr.*, 143, 314, 1977.
56. **Trefz, F. K., Byrd, D. J., and Kochen, W.**, Quantitative determination of cortisol in human plasma by high-pressure liquid chromatography, *J. Chromatogr.*, 107, 181, 1975.
57. **Schwedt, G., Bussemas, H. H., and Lippmann, Ch.**, Determination of free cortisol in urine by high-pressure liquid chromatography, *J. Chromatogr.*, 143, 259, 1977.
58. **Shaw, R. and Elliott, W. H.**, Bile acids XLVIII. Separation of conjugated bile acids by high-pressure liquid chromatography, *Anal. Biochem.*, 74, 273, 1976.
59. **Evans, J. E. and McCluer, R. H.**, High pressure liquid chromatography of neutral glycosphingolipids, *Biochim. Biophys. Acta*, 270, 565, 1972.
60. **Jungalwala, F. B., Evans, J. E., and McCluer, R. H.**, High-performance liquid chromatography of phosphatidylcholine and sphingomyelin with detection in the region of 200 nm, *Biochem. J.*, 155, 55, 1976.
61. **Carr, K., Sweetman, B. J., and Frolich, J. C.**, High performance liquid chromatography of prostaglandins: biological applications, *Prostaglandins*, 11, 3, 1976.
62. **Hubbard, W. C. and Watson, J. T.**, Determination of 15-keto-13,14-dihydrometabolites of PGE2 and PGF2 alpha in plasma using high performance liquid chromatography and gas chromatography-mass spectrometry, *Prostaglandins*, 12, 21, 1976.
63. **Möller, S. E.**, A method for the separation of plasma amino acids by ion-exchange chromatography combined with a semiautomatic computation of plasma amino acid concentrations, *J. Chromatogr.*, 104, 63, 1975.
64. **Ersser, R. S.**, Rapid ion-exchange chromatography of plasma aminoacids, in *High Pressure Liquid Chromatography in Clinical Chemistry*, Dixon, P. F., Gray, C. H., Lim, C. K., and Stoll, M. S., Eds., Academic Press, London 1976, 23.
65. **Adams, R. F., Schmidt, G. J., and Vandemark, F. L.**, Determination of ε-aminocaproic acid in serum by reversed-phase chromatography with fluroescence detection, *Clin. Chem.*, 23, 1226, 1977.
66. **Graffeo, A. P. and Karger, B. L.**, Analysis for indole compounds in urine by high-performance liquid chromatography with fluorometric detection, *Clin. Chem.*, 22, 184, 1976.
67. **Mee, T. J. X. and Smith, J. A.**, High pressure chromatographic separation of some biogenic amines and derivatives implicated in the aetiology of schizophrenia, in *High Pressure Liquid Chromatography in Clinical Chemistry*, Dixon, P. F., Gray, C. H., Lim, C. K., and Stoll, M. S., Eds., Academic Press, London, 1976, 119.
68. **Grushka, E., and Kikta, E. and Naylor, E.**, Tryptophan and kynurenine determination in untreated urine by reversed-phase high-pressure liquid chromatography, *J. Chromatogr.*, 143, 51, 1977.
69. **Trefz, F. K., Byrd, D. J., and Kochen, W.**, Point of micropreparative high pressure liquid chromatography and thin-layer chromatography for the identification of indole compounds in human plasma, *J. Clin. Chem. Clin. Biochem.*, 14, 65, 1976.
70. **Banda, P. W., Sherry, A. E., and Blois, M. S.**, Column cation-exchange separation of melanin-related metabolites in urine from cases of melanoma, *Clin. Chem.*, 23, 1397, 1977.
71. **Brown, N. D., Sing, H. C., Neeley, W. E., and Koetitz, S. E.**, Determination of "true" serum creatinine by high-performance liquid chromatography combined with a continuous-flow microanalyzer, *Clin. Chem.*, 23, 1281, 1977.
72. **Brown, P. R.**, The use of high-pressure liquid chromatography in pharmacology and toxicology, *Adv. Chromatogr.*, 12, 1, 1975.
73. **Brown, P. R.**, *High Pressure Liquid Chromatography Biochemical and Biomedical Applications*, Academic Press, New York, 1973.
74. **Scholar, E. M., Brown, P. R., Parks, R. E., and Calabresi, P.**, Nucleotide profiles of the formed elements of human blood determined by high-pressure liquid chromatography, *Blood*, 41, 927, 1973.
75. **Rao, G. H. R., White, J. G., Jachimowicz, A. A., and Witkop, C. J.**, Nucleotide profiles of normal and abnormal platelets by high-pressure liquid chromatography, *J. Lab. Clin. Med.*, 84, 839, 1974.
76. **Perrett, D.**, High performance liquid chromatography of nucleotides in biological fluids, in *High Pressure Liquid Chromatography in Clinical Chemistry* , Dixon, P. F., Gray, C. H., Lim, C. K., and Stoll, M. S., Eds., Academic Press, London, 1976, 109.
77. **Lis, A. W., McLaughlin, R. K., and McLaughlin, D. I.**, Urinary purine levels in suicides, *Physiol. Chem. Phys.*, 7, 325, 1975.
78. **Hartwick, R. A. and Brown, P. R.**, Selective analyses for adenosine using reversed-phase high-pressure liquid chromatography, *J. Chromatogr.*, 143, 383, 1977.
79. **Davis, G. E., Suits, R. D., Kuo, K. C., Gehrke, C. W., Waalkes, T. P., and Borek, E.**, High-performance liquid chromatographic separation and quantitation of nucleosides in urine and some other biological fluids, *Clin. Chem.*, 23, 1427, 1977.

80. **Mell, L. D. and Gustafson, A. B.**, Urinary free norepinephrine and dopamine determined by reversed-phase high-pressure liquid chromatography, *Clin. Chem.*, 23, 473, 1977.
81. **Molnar, I. and Horvath, C.**, Reverse-phase chromatography of polar biological substances: separation of catechol compounds by high-performance liquid chromatography, *Clin. Chem.*, 22, 1497, 1976.
82. **Knox, J. H. and Jurand, J.**, Separation of catecholamines and their metabolites by adsorption, ion-pair and soap chromatography, *J. Chromatogr.*, 125, 89, 1976.
83. **Jurand, J.**, Separation of catecholamines and their metabolites by high speed liquid chromatography, in *High Pressure Liquid Chromatography in Clinical Chemistry*, Dixon, P. F., Gray, C. H., Lim, C. K., and Stoll, M. S., Eds., Academic Press, London, 1976, 125.
84. **Bronaugh, R. L., Hattox, S. E., Hoehn, M. M., Murphy, R. C., and Rutledge, C. O.**, The separation and identification of dopamine 3-0-sulfate and dopamine 4-0-sulfate in urine of parkinsonian patients, *J. Pharmacol. Exp. Ther.*, 195, 441, 1975.
85. **Brown, N. D. and Michalski, E. J.**, An improved high-performance liquid chromatographic method for quantifying *p*-aminobenzoic acid and some of its metabolites, *J. Chromatogr.*, 121, 76, 1976.
86. **Brown, N. D., Lofberg, R. T., and Gibson, T. P.**, High-performance liquid chromatographic method for the determination of *p*-aminobenzoic acid and some of its metabolites, *J. Chromatogr.*, 99, 635, 1974.
87. **Terada, H., Hayaski, T., Kawai, S., and Ohno, T.**, High-performance liquid chromatographic determination of pyruvic acid and α-ketoglutaric acid in serum, *J. Chromatogr.*, 130, 281, 1977.
88. **Hayashi, T., Sugiura, T., Terada, H., Kawai, S., and Ohno, T.**, High-speed liquid chromatographic determination of phenylpyruvic acid, *J. Chromatogr.*, 118, 403, 1976.
89. **Yoshida, A., Yoshioka, M., Tanimura, T., and Tamura, Z.**, Determination of vanilmandelic acid and homovanillic acid in urine by high-speed liquid chromatography, *J. Chromatogr.*, 116, 240, 1976.
90. **Liao, J. C., Hoffman, N. E., Barboriak, J. J., and Roth, D. A.**, High-performance liquid chromatography of pyruvic and α-ketoglutaric acids and its application to urine samples, *Clin. Chem.*, 23, 802, 1977.
91. **Molnar, I. and Horvath, C.**, Rapid separation of urinary acids by high-performance liquid chromatography, *J. Chromatogr.*, 143, 391, 1977.
92. **Chang, S. H., Gooding, K. M., and Regnier, F. E.**, Use of oxiranes in the preparation of bonded phase supports, *J. Chromatogr.*, 120, 321, 1976.
93. **Chang, S. H., Noel, R., and Regnier, F. E.**, High speed ion exchange chromatography of proteins, *Anal. Chem.*, 48, 1839, 1976.
94. **Schroeder, R. R., Kudirka, P. J., and Toren, E. C.**, Enzyme - selective detector systems for high-pressure liquid chromatography, *J. Chromatogr.*, 134, 83, 1977.
95. **Kudirka, P. J., Schroeder, R. R., Hewitt, T. E., and Toren, E. C.**, High-pressure liquid-chromatographic separation of lactate dehydrogenase isoenzymes, *Clin. Chem.*, 22, 471, 1976.
96. **Pickart, L. R. and Thaler, M. M.**, Purification of growth-promoting peptides and proteins, and of histones, by high pressure silica gel chromatography, *Prep. Biochem.*, 5, 397, 1975.
97. **Slavin, W., Rhys Williams, A. T., and Adams, R. F.**, A fluorescence detector for high-pressure liquid chromatography, *J. Chromatogr.*, 134, 121, 1977.
98. **Hörchner, P. and Rietveld, T.**, Rapid method for the quantitative determination of porphyrin methyl esters by high-pressure liquid chromatography, *J. Chromatogr.*, 123, 414, 1976.
99. **Evans, N., Jackson, A. H., Matlin, S. A., and Towill, R.**, High-performance liquid chromatographic analysis of porphyrins in clinical materials, *J. Chromatogr.*, 125, 345, 1976.
100. **Evans, N., Jackson, A. H., Matlin, S. A., and Towill, R.**, HPLC analysis of porphyrins in biological materials, in *High Pressure Liquid Chromatography in Clinical Chemistry*, Dixon, P. F., Gray, C. H., Lim, C. K., and Stoll, M. S., Eds., Academic Press, London, 1976, 71.
101. **Gray, C. H., Lim, C. K., and Nicholson, D. C.**, Porphyrin profiles in porphyrias, in *High Pressure Liquid Chromatography in Clinical Chemistry*, Dixon, P. F., Gray, C. H., Lim, C. K., and Stoll, M. S., Eds., Academic Press, London, 1976, 79.
102. **Carlson, R. E. and Dolphin, D.**, Application of HPLC to the analysis of clinically important porphyrins, in *High Pressure Liquid Chromatography in Clinical Chemistry*, Dixon, P. F., Gray, C. H., Lim, C. K., and Stoll, M. S., Eds., Academic Press, London, 1976, 87.
103. **Pryde, A. and Darby, F. J.**, The analysis of paraquat in urine by high-speed liquid chromatography, *J. Chromatogr.*, 115, 107, 1975.
104. **Smalls, U. and Nipper, H. C.**, A High Performance Liquid Chromatographic Method for the Rapid Determination of Warfarin and Its Metabolites in Serum, presented at the 29th National Meeting, American Association for Clinical Chemistry, Chicago, July 1977.
105. **Wong, L. T., Solomonraj, G., and Thomas, B. H.**, Analysis of warfarin in plasma by high-pressure liquid chromatography, *J. Chromatogr.*, 135, 149, 1977.

106. **Twitchett, P. J., Gorvin, A. E. P., Moffat, A. C., Williams, P. L., and Sullivan, A. T.**, An evaluation of some HPLC columns for the identification and quantitation of drugs and metabolites, in *High Pressure Liquid Chromatography in Clinical Chemistry*, Dixon, P. F., Gray, C. H., Lim, C. K., and Stoll, M. S., Eds., Academic Press, London, 201, 1976.
107. **Jane, I. and Taylor, J. F.**, Characterisation and quantitation of morphine in urine using high-pressure liquid chromatography with fluorescence detection, *J. Chromatogr.*, 109, 37, 1975.
108. **Perchalski, R. J., Winefordner, J. D., and Wilder, B. J.**, Evaluation of Eimac lamp as excitation source for molecular fluorescence with application to quantitation of ergotamine in plasma by high-pressure liquid chromatography, *Anal. Chem.*, 47, 1993, 1975.
109. **Watson, I. D.**, Rapid analysis of nicotine and cotinine in the urine of smokers by isocratic high-performance liquid chromatography, *J. Chromatogr.*, 143, 203, 1977.
110. **Cartoni, G. P. and Natalizia, F.**, Determination of pemoline by high-pressure liquid chromatography, *J. Chromatogr.*, 123, 474, 1976.
111. **Tjaden, U. R., Kraak, J. C., and Huber, J. F. K.**, Rapid trace analysis of barbiturates in blood and saliva by high-pressure liquid chromatography, *J. Chromatogr.*, 143, 183, 1977.
112. **Dünges, W.**, High pressure liquid chromatographic analysis of barbiturates in the picomole range by fluorometry of their DANS-derivatives, *J. Chromatogr. Sci.*, 12, 655, 1974.
113. **Dixon, P. F. and Stoll, M. S.**, The HPLC detection of some drugs taken in overdose, in *High Pressure Liquid Chromatography in Clinical Chemistry*, Dixon, P. F., Gray, C. H., Lim, C. K., and Stoll, M. S., Eds., Academic Press, London, 1976, 165.
114. **Harzer, K. and Barchet, R.**, Analysis of benzodiazepines and their hydrolysis products, benzophenones, by reversed-phase high-performance liquid chromatography and its application to biological material, *J. Chromatogr.*, 132, 83, 1977.
115. **Bugge, A.**, Quantitative high-performance liquid chromatography of diazepam and N-desmethyldiazepam in blood, *J. Chromatogr.*, 128, 111, 1976.
116. **Kopjak, L., Jennison, T., McCloskey, K., and Peat, M.**, presented at the California Association of Toxicologists meeting, Los Angeles, January 1977.
117. **Moore, B., Nickless, G., Hallett, C., and Howard, A. G.**, Analysis of nitrazepam and its metabolites by high-pressure liquid chromatography, *J. Chromatogr.*, 137, 215, 1977.
118. **Vree, T. B., Lenselink, B., Van der Kleijn, E., and Nühuis, G. M. M.**, Determination of flunitrazepam in body fluids by means of high-performance liquid chromatography, *J. Chromatogr.*, 143, 530, 1977.
119. **Kabra, P. M. and Marton, L. J.**, Determination of carbamazepine in blood or plasma by high-pressure liquid chromatography, *Clin. Chem.*, 22, 1070, 1976.
120. **House, I. M. and Berry, D. J.**, Analysis of carbamazepine in plasma by high pressure liquid chromatography, in *High Pressure Liquid Chromatography in Clinical Chemistry*, Dixon, P. F., Gray, C. H., Lim, C. K., and Stoll, M. S., Eds., Aademic Press, London, 1976, 155.
121. **Adams, R. F., Schmidt, G. J., and Vandemark, F. L.**, A micro method for the determination of carbamazepine and the 10,11-epoxide metabolite in serum and urine by reverse-phase liquid chromatography, *Chromatogr. Newsl.*, 5, 11, 1977.
122. **Eichelbaum, M. and Bertilsson, L.**, Determination of carbamazepine and its epoxide metabolite in plasma by high-speed liquid chromatography, *J. Chromalogr.*, 103, 135, 1975.
123. **Kabra, P. M. and Marton, L. J.**, High-pressure liquid chromatographic determination of 5-(4-hydroxyphenyl)-5-phenylhydantoin in human urine, *Clin. Chem.*, 22, 1672, 1976.
124. **Hermansson, J. and Karlén, B.**, Assay of the major (4-hydroxylated) metabolite of diphenylhydantoin in human urine by reversed phase high-performance liquid chromatography, *J. Chromatogr.*, 130, 422, 1977.
125 . **Kabra, P. M.**, Simultaneous measurement of phenobarbital, diphenylhydantoin, and primidone in blood by high-pressure liquid chromatography, *Clin. Chem.*, 22, 824, 1976.
126. **Adams, R. F. and Vandemark, F. L.**, Simultaneous high-pressure liquid chromatographic determination of some anticonvulsants in serum, *Clin. Chem.*, 22, 25, 1976.
127. **Soldin, S. J. and Hill, J. G.**, Rapid micromethod for measuring anticonvulsant drugs in serum by high-performance liquid chromatography, *Clin. Chem.*, 22, 856, 1976.
128. **Adams, R. F. and Vandemark, F. L.**, Determination of anticonvulsants in serum by use of high pressure liquid chromatography, in *High Pressure Liquid Chromatography in Clinical Chemistry*, Dixon, P. F., Gray, C. H., Lim, C. K., and Stoll, M. S., Eds., Academic Press, London, 1976, 143.
129. **Westenberg, H. G. M. and Dezeeuw, R. A.**, Rapid and sensitive liquid chromatographic determination of carbamazepine suitable for use in monitoring multiple-drug anticonvulsant therapy, *J. Chromatogr.*, 118, 217, 1976.
130. **Kabra, P. M., Stafford, B. E., and Marton, L. J.**, Simultaneous measurement of phenobarbital, phenytoin, primidone, ethosuximide, and carbamazepine in serum by high-pressure liquid chromatography, *Clin. Chem.*, 23, 1284, 1977.

131. **Needham, L. L. and Kochhar, M. M.**, Determination of ethchlorvynol by high-pressure liquid chromatography, *J. Chromatogr.*, 111, 422, 1975.
132. **Mellström, B. and Eksborg, S.**, Determination of chlorimipramine and desmethylchlorimipramine in human plasma by ion-pair partition chromatography, *J. Chromatogr.*, 116, 475, 1976.
133. **Watson, I. D. and Stewart, M. J.**, Quantitative determination of amitriptyline and nortriptyline in plasma by high-performance liquid chromatography, *J. Chromatogr.*, 132, 155, 1977.
134. Determination of tricyclic antidepressants and metabolites in plasma, *Liquid Chromatography Applications. Biomedical Analysis,* Pye Unicam, Cambridge, 1976.
135. **Brodie, R. R., Chasseaud, L. F., and Hawkins, D. R.**, Separation and measurement of tricyclic antidepressant drugs in plasma by high-performance liquid chromatography, *J. Chromatogr.*, 143, 535, 1977.
136. **Kraak, J. C. and Bijster, P.**, Determination of amitriptyline and some of its metabolites in blood by high-pressure liquid chromatography, *J. Chromatogr.*, 143, 499, 1977.
137. **Rodgers, D. H.**,Separation and analysis of psychopharmacologic drugs by high efficiency liquid chromatography, *J. Chromatogr. Sci.*, 12, 742, 1974.
138. **Muusze, R. G. and Huber, J. F. K.**, Determination of the psychotropic drug thioridazine and its metabolites in blood by means of high pressure liquid chromatography in combination with fluorometric reaction detection, *J. Chromatogr. Sci.*, 12, 779, 1974.
139. **Watson, I. D. and Stewart, M. J.**, Assay of tricyclic structured drugs and their metabolites in urine by high-performance liquid chromatography, *J. Chromatogr.*, 134, 182, 1977.
140. **Weddle, O. H. and Mason, W. D.**, Rapid determination of theophylline in human plasma by high-pressure liquid chromatography, *J. Pharm. Sci.*, 65, 865, 1976.
141. **Smith, V. M., Jennison, T. A., and Kopjak, L.**, Liquid chromatography helps solve problems in toxicology, *Chromatogr. Rev.*, 3, 7, 1977.
142. **Hill, R. E.**, Retention behaviour of a bonded reversed phase in a high-performance liquid chromatographic assay of serum theophylline, *J. Chromatogr.*, 135, 419, 1977.
143. **Maijub, A. G. and Stafford, D. T.**, Theophylline and dyphylline levels in serum by liquid chromatography, *J. Chromatogr. Sci.*, 14, 521, 1976.
144. **Kelley, J. G. and Leahey, W. J.**, Measurement of theophylline in plasma by high performance liquid chromatography, *Br. J. Clin. Pharmacol.*, 3, 947, 1976.
145. **Thompson, R. D., Nagasawa, H. T., and Jenne, J. W.**, Determination of theophylline and its metabolites in human urine and serum by high-pressure liquid chromatography, *J. Lab. Clin. Med.*, 84, 584, 1974.
146. **Franconi, L. C., Hawk, G. L., Sandmann, B. J., and Haney, W. G.**, Determination of theophylline in plasma ultrafiltrate by reversed phase high pressure liquid chromatography, *Anal. Chem.*, 48, 372, 1976.
147. **Cooper, M. J., Mirkin, B. L., and Anders, M. W.**, A rapid micromethod for the high-performance liquid chromatographic determination of theophylline in human serum, *J. Chromatogr.*, 143, 324, 1977.
148. **Berg, J. and Abbott, S.**, Rapid analysis of theophylline in plasma by reverse phase liquid chromatography in, *Liquid Chromatography at Work,* Varian Aerograph, Walnut Creek, Calif., 1977, 34.
149. **Orcutt, J. J., Kozak, P. P., Gillman, S. A., and Cummins, L. H.**, Micro-scale method for theophylline in body fluids by reversed-phase high-pressure liquid chromatography, *Clin. Chem.*, 23, 599, 1977.
150. **Nelson, J. W., Cordry, A. L., Aron, C. G., and Bartell, R. A.**, Simplified micro-scale procedure for preparing samples for theophylline determination by liquid chromatography, *Clin. Chem.*, 23, 124, 1977.
151. **Sitar, D. S., Piafsky, K. M., Rangno, R. E., and Ogilvi R. I.**, Plasma theophylline concentrations measured by high-pressure liquid chromatography, *Clin. Chem.*, 21, 1774, 1975.
152. **Manion, C. V., Shoeman, D. W., and Azarnoff, D. L.**, High-pressure liquid chromatographic assay of theophylline in biological fluids, *J. Chromatogr.*, 101, 169, 1974.
153. **Abbott, S., Pecsar, R., and Berg, J.**, Analysis of theophylline in plasma in *Liquid Chromatography at Work,* Varian Aerograph, Walnut Creek, Calif., 1976, 26.
154. **Altmiller, D. H. and Fowler, M. W.**, Analysis of Salicylic Acid in Serum by Reverse Phase Liquid Chromatography, presented at the 29th National Meeting, American Association for Clinical Chemistry, Chicago, 1977.
155. **Blair, D. and Rumack, B. H.**, Acetaminophen in serum and plasma estimated by high-pressure liquid chromatography, *Clin. Chem.*, 23, 743, 1977.
156. **Wong, L. T., Solomonraj, G., and Thomas, B. H.**, High-pressure liquid chromatographic determination of acetaminophen in biological fluids, *J. Pharm. Sci.*, 65, 1064, 1976.
157. **Gotelli, G. R., Kabra, P. M., and Marton, L. J.**, Determination of acetaminophen and phenacetin in plasma by high-pressure liquid chromatography, *Clin. Chem.*, 23, 957, 1977.

158. **Rosano, T. G. and Meola, J. M.**, H.P.L.C. Identification of Common Analgesic Drugs with Quantitation of Acetaminophen, presented at the 29th National Meeting, American Association for Clinical Chemistry, Chicago, 1977.
159. **Horvitz, R. A. and Jatlow, P.**, A High Pressure Liquid Chromatographic Method for the Determination of Acetaminophen Concentrations in Serum, presented at the 29th National Meeting, American Association for Clinical Chemistry, Chicago, 1977.
160. **Horwitz, R. A. and Jatlow, P. I.**, Determination of acetaminophen concentrations in serum by high-pressure liquid chromatography, *Clin. Chem.*, 23, 1596, 1977.
161. **Skellern, G. G. and Salole, E. G.**, A high-speed liquid chromatographic analysis of indomethacin in plasma, *J. Chromatogr.*, 114, 483, 1975.
162. **Inaba, T., Besley M. E., and Chow, E. J.**, Determination of sulfinpyrazone in serum by high-performance liquid chromatography, *J. Chromatogr.*, 104, 165, 1975.
163. **Lecaillon, J.-B. and Souppart, C.**, Quantitative assay of sulphinpyrazone in plasma and urine by high performance liquid chromatography, *J. Chromatogr.*, 121, 227, 1976.
164. **Brown, M. and Bye, A.**, The determination of allopurinol and oxipurinol in human plasma and urine, *J. Chromatogr.*, 143 195, 1977.
165. **Dutcher, J. S. and Strong, J. M.**, Determination of plasma procainamide and N-acetylprocainamide concentration by high-pressure liquid chromatography, *Clin. Chem.*, 23, 1318, 1977.
166. **Shudur, L. R., Powers, J. L., Marques, R. A., Winter, M. E., and Sandee, W.**, Measurement of procainamide and N-acetylprocainamide in serum by high-performance liquid chromatography, *Clin. Chem.*, 23, 636, 1977.
167. **Carr, K., Woosley, R. L., and Oates, J. A.**, Simultaneous quantification of procainamide and N-acetylprocainamide with high-performance liquid chromatography, *J. Chromatogr.*, 129, 363, 1976.
168. **Rocco, R. M., Abbott, D. C., Giese, R. W., and Karger, B. L.**, Analysis for procainamide and N-acetylprocainamide in plasma or serum by high-performance liquid chromatography, *Clin. Chem.*, 23, 705, 1977.
169. **Adams, R. F., Vandemark, F. L., and Schmidt, G.**, The simultaneous determination of lidocaine and procainamide in serum by use of high pressure liquid chromatography, *Clin. Chim. Acta*, 69, 515, 1976.
170. **Meffin, P. J., Harapat S. R., and Harrison, D. C.**, High-pressure liquid chromatographic analysis of drugs in biological fluids. III. Analysis of disopyramide and its mono-N-dealkylated metabolite in plasma and urine, *J. Chromatogr.*, 132, 503, 1977.
171. **Meffin, P. J., Harapat, S. R., Yee, Y. G., and Harrison, D. C.**, High-pressure liquid chromatographic analysis of drugs in biological fluids. V. Analysis of acebutolol and its major metabolite, *J. Chromatogr.*, 138, 183, 1977.
172. **Graffeo, A. P., Lin, D. C. K., and Foltz, R. L.**, Analysis of benzoylecgonine in urine by high-performance liquid chromatography and gas chromatography-mass spectrometry, *J. Chromatogr.*, 126, 717, 1976.
173. **Needham, L. L. and Kochkar, M. M.**, Determination of ketamine and some *in vivo* metabolites using high-pressure liquid chromatography, *J. Chromatogr.*, 114, 220, 1975.
174. **Magic, S. E.**, Determination of β-cetotetrine in plasma and urine using high-performance liquid chromatography with electrochemical detection, *J. Chromatogr.*, 129, 73, 1976.
175. **Cooper, M. J. Anders, M. W., and Mirkin, B. L.**, Ion-pair extraction and high-speed liquid chromatography of cephalothin and diacetylcephalothin in human serum and urine, *Drug Metab. Dispos.*, 1, 659, 1973.
176. **Buhs, R. P., Maxim, T. E., Allen, N., Jacob, T. A., and Wolf, F. J.**, Analysis of cefoxitin, cephalothin and their deacylated metabolites in human urine by high-performance liquid chromatography, *J. Chromatogr.*, 99, 609, 1974.
177. **Dell, D.**, The determination of the semi-synthetic cephalosporin HR580 in plasma and urine, in *High Pressure Liquid Chromatography in Clinical Chemistry*, Dixon, P. F., Gray, C. H., Lim, C. K., and Stoll, M. S., Eds., Academic Press, London, 1976, 185.
178. **Hulhoven, R. and Desager, J. P.**, Quantitative determination of low levels of daunomycin and daunomycinol in plasma by high-performance liquid chromatography, *J. Chromatogr.*, 125, 369, 1976.
179. **Benvenuto, J. A., Lu, K., and Loo, T. L.**, High-pressure liquid chromatographic analysis of ftorafur and its metabolites in biological fluids, *J. Chromatogr.*, 134, 219, 1977.
180. **Shargel, L., Koss, R. F., Crain, A. V. R., and Boyle, V. J.**, Nalidixic acid and hydroxynalidixic acid analysis in human plasma and urine by liquid chromatography, *J. Pharm. Sci.*, 62, 1452, 1973.
181. **Murray, J. F., Gordon, G. R., Gulledge, C. C., and Peters, J. H.**, Chromatographic-fluorometric analysis of antileprotic sulfones, *J. Chromatogr.*, 107, 67, 1975.
182. **Papp, E., Magyar, K., and Schwarz, H. J.**, Liquid-solid chromatographic determination of 6-demethylgriseofulvin in urine, *J. Pharm. Sci.*, 65, 441, 1976.

183. **Weber, D. J.**, High-pressure liquid chromatographic analysis of tolbutamide in serum, *J. Pharm. Sci.*, 65, 1502, 1976.
184. **Sved, S., McGilveray, I. J., and Beaudoin, N.**, Assay of sulfonylureas in human plasma by high-performance liquid chromatography, *J. Pharm. Sci.*, 65, 1356, 1976.
185. **Ross, M. S. F.**, Determination of metformin in biological fluids by derivatization followed by high-performance liquid chromatography, *J. Chromatogr.*, 133, 408, 1977.
186. **Little, C. J., Dale, A. D., Ord, D. A., and Marten, T. R.**, Determination of mefruside and its metabolites in urine by high performance liquid chromatography, *Anal. Chem.*, 49, 1311, 1977.
187. **Bjornsson, T. D., Blaschke, T. F., and Meffin, P. J.**, High-pressure liquid chromatographic analysis of drugs in biological fluids. IV. Determination of clofibrinic acid, *J. Chromatogr.*, 137, 145, 1977.
188. **Woodhouse, R. N., Cresswell, D. G., Chasseaud, L. F., and Brodie, R. R.**, Use of high-performance liquid chromatography to measure plasma concentrations of *p*-chlorophenoxyisobutyric acid after administration of clofibrate to humans, *J. Chromatogr.*, 137, 218, 1977.
189. **Christophersen, A. S., Rasmussen, K. E., and Salvesen, B.**, Determination of hydrochlorothiazide in serum by high-pressure liquid chromatography, *J. Chromatogr.*, 132, 91, 1977.
191. **Cooper, M. J., Sinaiko, A. R., Anders, M. W., and Mirkin, B. L.**, High Pressure liquid chromatographic determination of hydrochlorothiazide in human serum and urine, *Anal. Chem.*, 48, 1110, 1976.
192. **Cooper, M. J., Anders, M. W., Sinaiko, A. R., and Mirkin, B. L.**, Application of HPLC to the study of the disposition of hydrochlorothiazide in adults and children, in *High Pressure Liquid Chromatography in Clinical Chemistry*, Dixon, P. F., Gray, C. H., Lim, C. K. and Stoll, M. S., Eds., Academic Press, London, 1976, 175.
193. **Skellern, G. G., Knight, B. I., and Stenlake, J. B.**, Improved method for the determination of methimazole in plasma by high-performance liquid chromatography, *J. Chromatogr.*, 124, 405, 1976.
194. **Roggia, S., Grossoni, G., Pelizza, G., Ratti, B., and Gallo, G.**, Determination of diftalone and its main metabolites in human plasma and synovial fluid by high-performance liquid chromatography, *J. Chromatogr.*, 124, 169, 1976.
195. **Watson, I. D.**, personal communication.

AUTOMATED MICROBIOLOGY

W. Scott Ramsey

TABLE OF CONTENTS

I. Introduction 104

II. The Challenge of Microbiology to Automation 104
A. Detection of Significant Microbes 105
B. Identification of Microbes 106
C. Determination of Antimicrobial Susceptibilities 108
1. Disk Diffusion 108
2. MIC 109

III. Approaches to Automation 109
A. Radiometric Detection 109
B. Impedance 111
C. Selective Growth 112
D. Scattered Light Spectra 113
E. Light Scattering 114
F. Optical Density 115
G. Plate Streaking 116
H. Microcalorimetry 116
I. ATP Detection 117
J. Gas Chromatography 118
K. Specific Enzymatic Profiles 118
L. Fluorescent Spectrograms 119
M. Particle Analysis 120
N. Electrophoresis 120

IV. Laboratory Considerations 120
A. Reasons for Automation 120
B. Financial Analysis 120
C. Purchase Considerations 121
D. Federal Regulation 121

V. Future Projections 122

References 122

I. INTRODUCTION

It is frequently asked why automation has not reached the microbiology laboratory. The answer often given is that the responsibilities and procedures of the microbiology laboratory are so diverse and individualized that the standardization necessary to make automation worthwhile cannot exist. In a related view, microbes are held to be unsuitable for mechanized treatment by virtue of being alive and therefore inherently variable. It is also very difficult to imagine an automated system capable of handling all possible microbes without human intervention — rare organisms do exist and must be dealt with regardless of cost/benefit factors. It is true that the number of different reports which may issue from a microbiology laboratory is larger than from a clinical chemistry service. Nevertheless, the theme of this article will be that automation has arrived at the microbiology laboratory and may be expected to have a major impact in the near future.

It is rather difficult to follow automated microbiology in the traditional scientific literature. The main reason is that most devices and approaches are quite recent and studies performed with them simply have not appeared in the literature. Most references are in the 1970s.

There also is a tendency for manufacturers to introduce products at symposia and meetings of scientific societies. These presentations subsequently appear as abstracts or proceedings and are less complete, less accessible, and less likely to appear in abstract media than papers in traditional journals. Fortunately, an excellent bibliography has been prepared by Palmer and LeQuesne which covers the field from 1967 to 1975.[1] In addition, there are several proceedings of conferences.[2-6] A review by Isenberg and MacLowry[7] provides an excellent summary of the field. I will not attempt to cover all available papers. The format will be

1. To discuss the challenge to automation presented by microbiology
2. To present the approaches which appear to be most promising, with emphasis on commercially available devices
3. To discuss points to consider before acquiring automated instruments
4. To attempt to project the future of automated microbiology

II. THE CHALLENGE OF MICROBIOLOGY TO AUTOMATION

The duties of the clinical microbiology laboratory may be summarized by three questions:

1. In view of the patient's condition, are microbes of clinical significance in the specimen?
2. If so, what is the identification of the microbe or microbes?
3. In many cases, what are the antimicrobial susceptibilities of the microbes?

These questions are not posed explicitly, and many procedures answer more than one question. For example, a physician who makes a throat swab is interested primarily in the presence or absence of Group A streptococci. To test for this pathogen, one procedure is to streak the swab on a blood agar plate which is incubated overnight. The next day, Group A streptococci are distinguished from other β-hemolytic streptococci by streaking an isolated colony on a fresh plate and including on the plate a disc containing a low level of bacitracin. Most strains of Group A streptococci are sensitive to low levels of bacitracin, and a zone of inhibition which appears the following day allows a presumptive report to be made. (The antimicrobial susceptibilities for this organism are well known.) Should the report also include the presence or absence of normal throat flora? Should one risk missing some Group A streptococci, but gain a day by putting the bacitracin disk on the primary isolation plate (a procedure which is not recommended)?[8] The answers to these questions vary and depend on the prevalence of disease in the population, the resources of the laboratory, and on the needs of the physicians. This simple example which raises difficult questions may serve as an introduction to consideration of the principal duties of the microbiology laboratory.

A. Detection of Significant Microbes

How good should a system be at detecting microbes while avoiding false positive reports? What are the hazards of missing an infection and making a false negative report? In considering these questions, I will borrow heavily from Galen and Gambino's book.[9] In evaluating an assay one should consider the sensitivity: the ability of the assay to recognize positive samples

$$\% \text{ Sensitivity} = \frac{\text{detected positives}}{\text{true positives}} \times 100\text{; and the specificity,}$$

and the specificity, the ability of the assay to recognize negative samples

$$\% \text{ Specificity} = \frac{\text{detected negatives}}{\text{true negatives}} \times 100$$

Galen and Gambino also emphasize consideration of predictive values, defined as:

$$\% \text{ Predictive value of positive test} = \frac{\text{true positives}}{\text{total detected positives}} \times 100$$

The final factor which affects the specifications for a test is the prevalence of the disease — the number of cases in the population being tested. As we shall see, the prevalence has a profound effect on the capabilities of an assay. These points were illustrated in considering the specifications for a hypothetical screening assay to detect gonorrhea. The population aged 20 to 24 had a prevalence of the disease of 2,000/100,000 in 1972. An assay of 100% sensitivity and 95% specificity applied to 100,000 people aged 20 to 24 years would have resulted in 6,900 positive reports. Of these, 2,000 would have been true positives, and 4,900 would have been false positives. There would have been more than twice as many false positive as true positive. Thus, the predictive value of a positive report would be

$$\frac{2000}{6900} \times 100 = 29\% \text{ and the predictive value of a negative report would be}$$

$$\frac{98,000}{93,100} \times 100 = 105\%$$

$$\text{The efficiency would be } \frac{2000 + 93,100}{100,000} = 95\%$$

A specificity of 99% would result in a predictive value of a positive result of only 67% — the predictive value of a positive result would reach 100% only as specificity achieved 99.9%. (One may note parenthetically that in this population, an assay which never gave a positive response no matter what would have a sensitivity of 0%, specificity of 100%, and efficiency of 98%).

What does all this mean in evaluating an assay? The prime lesson is that sensitivity and specificity figures for each assay must be evaluated in terms of prevalence, the risk of false negatives, and the risk and cost of false positives. Sensitivity and specificity have little meaning without knowledge of prevalence. Finally, sensitivity becomes of prime importance in high-prevalence populations and specificity of prime importance in low-prevalence populations.

B. Identification of the Microbes

Not all laboratories use the same criteria to identify microbes. Put another way, all labs use the same tests in making identifications, but some use more tests than others. For example, many laboratories report as "organisms resembling *Neisseria gonorrhoeae*" all Gram-negative oxidase-positive diplococci isolated on Thayer-Martin medium from a cervical swab. There is a small chance that these bacteria are not *N. gonorrhoeae* at all, but other *Neisseria* species. The identification may be confirmed by sugar fermentation tests which require at least one additional day for completion. Both reports are *de facto N. gonorrhoeae,* but a lab which did the sugar tests has a higher degree of confidence in this identification. There is a tendency to strive for the higher confidence identification. Yet, as Bartlett[10] emphasizes, the laboratory should aspire to the efficient generation of information of optimum value when delivered to the physician. Perhaps the information obtained by doing additional tests is of lesser value because of the time required. The development and evaluation of automated methods require a high degree of agreement between laboratories on the criteria used in making identifications. The use of prepared media and identification kits has gone a long way toward achieving this agreement. Further standardization ideally will be achieved through collaboration between clinicians and microbiologists.

Microbes classically have been identified in the clinical laboratory using an invariant progression of key characteristics; e.g., an isolate stained Gram positive or negative produced acid from lactose or did not and did or did not produce hydrogen sulfide from protein hydrolysates. Unfortunately, while isolates may be characterized in these terms, species generally may not. Only 90.8% of *Escherichia coli* strains produce gas from lactose; of *Proteus vulgaris,* 98.2% give a positive indole reaction. In a monumental study, Edwards and Ewing characterized the Enterobacteriaceae in terms of percentages of strains with specific biochemical responses.[11] Since most of the biochemical reactions used to identify microbes are subjct to this variation, the microbiologist must make decisions under conditions of uncertainty.

The method in common use for making decisions under conditions of uncertainty is based on Bayes' theorem.[12] We are interested in the problem expressed in this question: If 0 occurs (observed set of biochemical reactions of an isolate), what is the probability that the isolate is a member of the species A (or A_1 or A_2, etc.)?

In the general form, Bayes' theorem states:

$$P(A/O) = \frac{P(O \text{ and } A)}{P(O)}$$

when P(A) = probability that A occurs, P(O) = probability that O occurs and P (A/O) = probability that A occurs when 0 occurs. P(O/A) = probability that 0 occurs when A occurs. P(A) and P(O) are prior probabilities, i.e., they do not depend on other conditions. P(A/O) and P(O/A) are conditional probabilities in that they depend on the occurence of 0 and A, respectively. When this approach is used to identify

Table 1
INCIDENCE OF POSITIVE RESULTS IN GROWTH TESTS

	Test			
Species	**1**	**2**	**3**	**4**
A	0.98	0.01	0.5	0.6
B	0.5	0.98	0.01	0.4
C	0.01	0.98	0.98	0.5
Unknown	+	+	−	+

Probability unknown is species

A (0.98) (0.01) (0.5) (0.6) = 0.00294
B (0.5) (0.98) (0.99) (0.4) = 0.19404
C (0.01) (0.98) (0.02) (0.5) = 0.000098

Normalized probabilities unknown is species

A 0.00294/0.197029 = 0.0149
B 0.19404/0.197029 = 0.985
C 0.000098/0.197029 = 0.000497

microbes, all prior probabilities are assumed to be equal. Either quantitative or qualitative test results may be used in the set of observed reactions. To illustrate the use of Bayes' theorem, let us assume one wishes to compare an unknown isolate to three known species. Four growth tests have been used; the incidence of positive results in the known species and the growth patterns of the isolate is known (see Table 1).

Although the isolate growth pattern did not exactly match that for the majority of isolates of any species, the greatest probability is that the unknown belongs in species B. This is emphasized by normalizing the probabilities by dividing each probability by the sum of all probabilities. In this illustration, the various tests do not contribute to the identification to the same degree. Test 4 has little effect on the outcome whether it is positive or negative. It is also necessary to make an arbitrary decision concerning the level of probability one will accept as indicating identification — is it enough that the probability be highest of all species in the universe? What if the isolate has a very low probability of belonging in any of the three species?

This concept of identification was applied in a study of 70 species of Gram-negative rod-shaped bacteria.[13-15] A total of 50 different tests based mainly on biochemical reactions were used. The tests were qualitative. Of 1079 reference strains and 516 field strains presented as unknowns, the probabilistic method provided the correct identification for 90.8% of the fermentative and 82.1% of the nonfermentative reference strains. The correct identifications for the unknown strains were 89.4 and 70.8%, respectively. Of the 51 species not in the system, 20% were incorrectly identified as in the scheme. The success rate on the unknowns may have been unusually low because of rare strains.

This study is of particular value for the insights provided on practical problems of identification by probabilistic methods. One of the most vexing problems was in the generation of probability tables for the species based on data in the scientific literature. "Always" and "never" were given values of 0.99 and 0.01, respectively. "Usually" became 0.90 or 0.85. Any practical automated device using identification by probabilities will require the generation of test information from "scratch" using the actual tests. In this study, the probabilities were normalized, and an identification was considered firm, only if a certain level of normalized probability was reached. Lower values evoked suggestions for additional tests. This practice was thought to minimize the chance of a misidentification because of a single incorrect laboratory result. It is important that only one of the linked results be included in the calculations. The most efficient individual tests are those which tend to divide the universe of species in two equal parts. Tests may be ranked in efficiency by multiplying the number of species giving a positive response by the number giving a negative response. One valuable emperical finding was that at least two nonoverlapping (0.99/0.01) tests were required to separate species from their nearest neighbors reliably.

The first application of a probabilistic method of identification to bacteria was made by Dybowski and Franklin,[16] who studied the enterobacteria. Gyllenberg and Niemela[17] have studied the basic problem of deciding how high a probability must be for a confident identification to be made. They divided the results into four states, depending on whether the probabilities exceed predefined absolute and normalized probability limits.

When automated identification systems are compared to conventional methods, the results are generally expressed in terms of percentage correct results. Sensitivity and specificity may be used. The factor of prevalence is always important. Some laboratories find one third of their isolates to be *E. coli.* An identification system tested against this population would do well to have a high sensitivity and specificity for *E. coli.*

C. Determination of Antimicrobial Susceptibilities

The third major responsibility of the clinical microbiology laboratory is to provide information on the antimicrobial susceptibility of isolates when required. The idea is sometimes expressed that the physician really needs to know the antimicrobial susceptibility rather than the identification in order to treat patients. We shall see that this is not actually the case, although it conceivably could become the case in the future.

The pattern of antimicrobial susceptibilities is both well known and relatively invariant for many microbes. For example, the isolation of Group A streptococci, pathogenic *Neisseria,* or *Streptococcus pneumoniae* provides enough information for treatment. Antimicrobial susceptibilities are not done on these species. (The emergence of penicillin-resistant strains may cause a change in this practice for *N. gonorrhoeae.*)

A microbe is said to be sensitive or resistant to an antimicrobial based on its growth in vitro in the presence of the antimicrobial agent and on the level of drug found in a patient's serum after a reasonable dose. Admittedly there are many factors which affect the efficiency of drugs, such as individual variations in serum levels, effects of host defense mechanisms on drug action, and the relation between drug level at the site of infection and serum level. Still, the serum level provides a practical means of assessing sensitivity. The in vitro effect of a drug on a microbe is expressed as the minimum inhibitory concentration (MIC) or that concentration which just inhibits growth. An isolate is regarded as sensitive if the MIC is exceeded three to five times by the expected serum or body fluid drug level. An isolate whose MIC is not clearly exceeded by serum drug levels would be called resistant.[18]

1. Disk Diffusion

The standard method for determining antimicrobial susceptibility is a plate disk-diffusion method referred to as Kirby-Bauer (or Bauer-Kirby). Paper disks containing dried antimicrobial agents are placed on an agar plate seeded with the isolate. After overnight incubation, the diameter of the zone of growth inhibition allows the isolate to be characterized as sensitive, intermediate, or resistant with respect to each antimicrobial. In a carefully standardized test, the diameter of the inhibition zone is related to the log of the MIC. The pertinent zone sizes vary with each antimicrobial. In addition, it is important to note that the disk-diffusion method was developed for us with fresh isolates of rapidly growing, commonly isolated, aerobes, i.e., the Enterobacteriaceae, *Staphylococcus aureus,* and *Pseudomones aeruginosa.*[19] The correct interpretation of Kirby-Bauer zones for many antibiotics requires identification of the microbes.

2. MIC

It is generally thought desirable to directly determine the MIC for an isolate/antibiotic rather than express the sensitivity in terms of sensitive-intermediate-resistant. MIC determination has been recommended in certain clinical situations including endocarditis, other serious Group D streptococci infections, infections with slow-growing or fastidious organisms, unusual CNS infections, recurrent infections with the same pathogens, or infections in which the disk-diffusion method is inadequate.[18] MIC determinations involve growing the organism in a series of 1 plus 1 dilutions of each antimicrobial. The MIC is that concentration at which little or no growth occurs. This procedure was first developed using liquid media, but comparable results are obtained using antimicrobial dilutions in solid media.[20] Although more useful information about an isolate is contained in the MIC rather than the sensitive-intermediate-resistant determinations, considerably more technician time and materials are involved in a MIC. It is a situation that calls for automation, and several devices seek to meet this challenge.[21]

In this discussion, I have neglected the scrupulous attention to detail required to produce reliable antimicrobial susceptibility and MIC results. When new techniques are compared to the disk-diffusion method, the results are generally expressed in percentage agreement with the Kirby-Bauer categories. Disagreements are often subdivided as being minor (one test giving intermediate and the other sensitive or resistant) or major (with one assay giving sensitive and the other resistant). The most serious disagreement is when the test assay finds an isolate sensitive and the Kirby-Bauer method finds it resistant. When comparing methods for determining MICs, in at least one case, a MIC method was considered reliable when at least 95% of the replicate results fell within a single dilution of the MIC determined by the reference method.[22]

III. APPROACHES TO AUTOMATION

The first step is to limit the scope of this discussion. Virtually all approaches to automation of clinical microbiology have been directed toward bacteriology, as opposed to virology or mycology, presumably because of the greater demand for bacteriological services. The automation of serological and immunological procedures will not be included because the basic problems are quite different. Rapid methods of determining biochemical profiles such as R/B®, API®, Patho Tec®, and Enterotube® will not be included, primarily because these are now well-established procedures.

A. Radiometric Detection

Clinical microbiology has benefited from several spin-offs of the space exploration program. The detection of biological activity through assay of volatile $^{14}CO_2$ produced from labeled compounds was developed to detect life elsewhere in the solar system. It is one of the three life detection systems on the Viking Mars landers. An interim report on this experiment concludes: "Despite all hypotheses to the contrary, the distinct possibility remains that biological activity has been observed on Mars."[23]

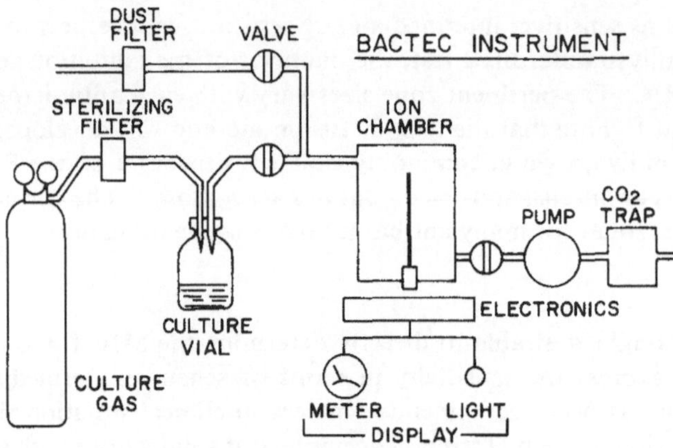

FIGURE 1. Schematic of Bactec® instrument. A culture vial containing bacteria, radioactive substrate, and $^{14}CO_2$ in the headspace is punctured with two sterile needles. Gas in the vial is flushed out with sterile gas and enters the ionization chamber where the level of radioactivity is determined. Finally, the ionization chamber is flushed into a trap, and the needles are withdrawn and resterilized. (From Bactec® brochure, Johnston Laboratories, Inc., Cockeysville, Md. With permission.)

Bacteremia may be detected by incubating blood with media containing ^{14}C-substrates in a sealed bottle followed by assay for $^{14}CO_2$ in the headspace.[24,25] Instruments and media are produced by Johnston Laboratories Inc. (Cockeysville, Md.). Figure 1 is a schematic of the assay process. Currently 5 models are produced, ranging from a manually operated device which only samples the headspace gas to an instrument which incubates the bottles at a preset temperature, stirs each bottle, and automatically samples up to 25 bottles. Each model sterilizes the sampling needles after use. Bactec® use has been extended to include spinal, synovial, and any other normally sterile fluid. Two media are produced for bacteremia detection. Both are essentially enriched soy digest broth, but one is gassed with 5% CO_2 in air, and the other is prereduced and gassed with a mixture of N_2 and CO_2. The essential feature of these media is a proprietary mixture of ^{14}C substrates amounting to about 1.5 μCi per vial. This amount of radioactive material is so low that no federal license is required, nor are any restrictions on storage, use, or disposal. A chronic problem has been the occurrence of false positive tests from $^{14}CO_2$ produced by blood elements, probably leukocytes. Sodium polyanetholesulfonate is included in each media to counter this effect.

A number of laboratories have reported on clinical evaluation of the Bactec® technology.[26,27] One study[28] extending over several years at two institutions will be considered in some detail. Blood samples were taken from patients in two 10-mℓ tubes containing polyanetholesulfonate. Aerobic and anaerobic Bactec® bottles were each inoculated with 3 mℓ of sample. Exactly 5 mℓ was added to 50-mℓ bottles containing tryptic soy broth or thioglycollate media. Aerobic Bactec® bottles were read at intervals from 1 to 4 hr the first day and on the third and seventh days. Anaerobic bottles were read daily. Subcultures were made after positive readings and 7 days. Conventional bottles were subcultured the morning after inoculation, after positive appearance, and after 7 and 14 days. A total of 1530 blood samples yielded 144 positive cultures and 137 contaminants. Of the total positive cultures, Bactec® detected 84%, and the conventional method detected 75%. A total of 25 false positives were reported by Bactec®. A total of 17% of the Bactec® positives were false positives (excluding contaminants — no data was given on relative detection of contaminants.) The conventional method detected most positives on the second day after arrival. Further studies using Bactec® alone showed that 32% of aerobic organisms were detected by 12 hr, 51% by 24 hr, and 82% by 48 hr. A total of 21% of the anaerobes were detected by the first day, 38% by the second day, and 68% by the third day. Some isolates were detected only by Bactec®; others were detected only by convenional methods. It should be noted, however, that quite different media were used with each method. The general conclusion was that Bactec® allowed the detection of significant bacteremia quicker than conventional methods.

Two specialized media kits extend Bactec's® versatility. One kit is a set of three vials containing ^{14}C-glucose, ^{14}C-maltose, or ^{14}C-fructose and ONPG discs. These are used in a 3-hr differentiation of *Neisseria* species. Another new kit is designed to provide rapid analysis of aminoglycosides in patient sera. It is based on the fact that gentamycin inhibits *Proteus* conversion of ^{14}C-urea to $^{14}CO_2$.

Schrot et al.[29] have shown that bacteria may be identified in approximately 1 hr by means of the species-specific rates of release of $^{14}CO_2$ from single labeled compounds. They developed a method in which a culture is incubated in a variety of substrates and the CO_2 produced is captured by barium hydroxide-soaked pads. Further development of this process required the generation of a "library" of profiles for 30 or more strains for each of 20 species. This process was also adapted to the rapid determination of antimicrobial susceptibilities.[30]

Burgeois and Mafart[31] used a reversal of this method to detect viable cells. The fluid to be assayed was filtered through a membrane filter which was incubated on a pad containing ^{14}C-lysine. After washing, the filters were assayed in a scintillation counter. A total of 2000 yeast cells could be detected after a 1-hr incubation.

B. Impedance

In 1899, Steward reported an increase in the conductivity of blood on putrification.[32] Little was done with this and subsequent observations until the 1970s, when Shatton and Company, Ltd. (Hartfield, Hertshire, England)[33] and Bactomatic Inc., (Palo Alto, Calif.)[34] produced devices for measuring of microbial-induced impedance changes. Impedance is the resistance to flow of an alternating current. It has resistance and reactive components and may be thought of as a resistor and capacitor in series.[35] It is difficult to accurately follow impedance in a single vessel containing fluid over a period of time because temperature variations and evaporation affect this parameter. In order to avoid this problem, the material containing bacteria and a sterile reference medium are run in parallel, and the difference in impedance between the two recorded. Growth is generally accompanied by a decrease in impedance, although an increase is sometimes seen.

The Stratton and Company Bactobridge uses cylindrical cuvettes with electrodes at either end. The entire cuvette is filled with medium and the entire contents subjected to the current. *E. coli* at 10^4 cells per milliliter could be detected with this instrument using a cuvette of 0.1 mℓ capacity. Larger cuvettes are planned for use in examining blood cultures.[36]

The Bactomatic Inc. instruments, Bactometer 8® and Bactometer 32®, use two types of detectors. In one configuration, a series of cylindrical vessels are cemented to a circuit board above printed electrodes. The circuit board fits into the instrument. Bactometers® also may be used with conventional blood culture bottles using electrodes attached to bottle caps. The Bactometer 8® accepts eight samples (and eight reference media) and is designed to fit in an incubator. The Bactometer 32® takes 32 samples (and 32 references) and has a self-contained incubator. Each model reports the changes in impedance periodically, on demand, or on an optional strip chart. Although these instruments are in clinical trials, some data concerning clinical use of these instruments are available. The circuit board arrangement of sample wells has been used to screen urines for significant bacterial levels. The threshold level for detection of bacteria by Bactometers® is 10^6 to 10^7 cells per milliliter. Urine specimens containing more than 10^5 cells per milliliter reached this threshold with 3 hr of incubation, while lower bacterial concentrations took significantly longer.[37] Parallel conventional and Bactometer 32® tests have been reported for 200 urine specimens.[38] A total of 41 specimens with clinically significant bacteriuria (here defined as 10^4 to 10^5 cells per milliliter) were detected by 2.5 hr, and 39 specimens with fewer than 10^4 cells per milliliter were detected in more than 5 hr. Approximately 4% of the organisms detected by conventional means were missed by Bactometer 32®, and 2% of the impedance positive specimens were negative by the conventional method. Bactometer® and conventional methods were compared in the examination of 1271 blood samples.[39] Microorganisms were isolated from 161 cultures. In 9 cases, there were impedance changes without microbial growth, and in 16 cases, no impedance changes occurred, although growth was observed. In a comparison of Bactometer 32® and conventional methods on 500 blood speciments, 15% were positive with both methods, and 14.5% were positive by impedance alone. The average detection time was 8.5 hr compared to 24 hr for the conventional method. The same authors reported the Bactometer® useful for early detection of *Haemophilus influenzae* in the cerebrospinal fluid of children.[40]

The Bactometer® module may be monitored while incubating in an anaerobic chamber. This technique has been suggested as a means of doing rapid antimicrobial susceptibility testing of anaerobes.[41]

In a somewhat different approach to the use of impedance to detect bacteria, Goldschmidt and Wheeler[42,43] described a cell using four in-line electrodes. The outer two electrodes provide the current that is sensed by the inner electrode pair. This instrument measures the effect on impedance of microbes suspended in distilled water rather than the metabolic effects of microbial growth. The relation between voltage output and frequency is affected by the presence of microbes, and 10^3 cells per milliliter could be detected. A total of 40% of 60 clinical urine samples contained fewer than 10^3 cells per milliliter by this and conventional means, while a 90 to 95% correlation between methods was reported for the samples containing more than 10^3 cells per milliliter.

C. Selective Growth

The McDonnell Douglas (St. Louis, Mo.)-Fisher Scientific Company (Chicago) Auto-Microbic System® (AMS)® is the most ambitious contribution to automated microbiology yet. A McDonnell Douglas subsidiary, Vitek Systems, Inc., has been set up to produce this system. It is designed with the potential of detecting significant infections, identifying the pathogens present, and determining antimicrobial susceptibilities. The initial effort has been directed toward urine samples.

The heart of this system is a disposable plastic "microplate" containing 20 interconnected wells, each of which can contain one freeze-dried medium. The procedure for analysis of a urine sample is as follows. Urine samples of 0.2 and 0.005 m*l* are pipetted into the disposable sample injector which is attached to the microplate. Diluent is automatically added by a diluent dispenser module. A rack of plates is placed in the filling module where the microplate chambers are automatically filled with diluted sample. Sample injectors are removed and discarded, and the microplate is placed in the reader/incubator module. Each module holds 120 microplates, and two modules may be combined with one computer/control for a capacity of 240 tests per 13-hr period. Every plate is read hourly for 12 hr, and results in terms of (1) significant bacteriuria (defined as exceeding 7×10^4 cells per milliliter) and (2) the identification of the 9 most common urine pathogens are displayed and printed. The identifications reported are *Pseudomonas aeruginosa, Proteus* spp., *Citrobacter freundii, Serratia*

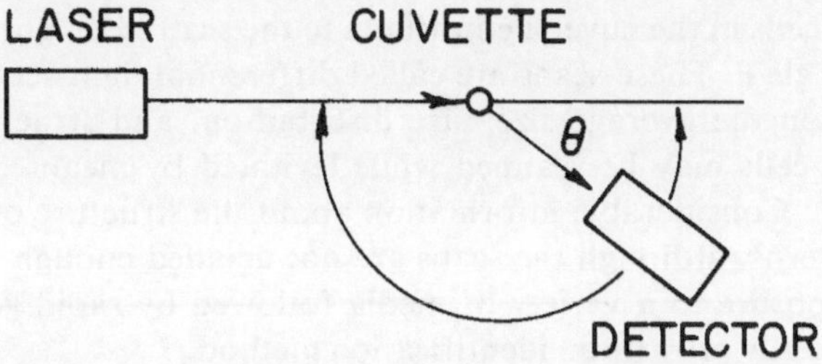

FIGURE 2. Schematic of differential light scattering determination. The laser source illuminates a suspension of cells in a cuvette. The detector measures the scattered light at angle θ as it rotates about the cuvette. (From Differential III® brochure, Science Spectrum, Inc., Santa Barbara, Calif. With permission.)

spp., *E. coli, Klebsiella/Enterobacter,* yeasts, *Enterococcus* Group D, and *Staphylococcus aureus.*[44]

The key to the AMS® is the development of specific media which indicate growth by color changes, precipitation reactions, or turbidity. One of the most important aspects of this system is the capability to identify components of mixed cultures. One well is designed to allow growth of organisms other than the nine most common ones. Over 3500 seeded urines were tested with ≥ 90% agreement with conventional methods. Also, 2000 clinical specimens provided similar results, except for enterococci and staphylococci (85 to 90% agreement).[45]

Sonnenwirth[46] tested a prototype instrument and found that agreement in identifying seeded specimens by AMS® and conventional methods ranged from 82 (*Citrobacter*) to 98% (*E. coli*). In clinical specimens, agreement was ≥90% overall and ranged from 100% for *Klebsiella/Enterobacter* to 79% for *P. aeruginosa.* Agreement on negative specimens was at least 94%.

A collaborative study involving six laboratories compared identification of seeded pure and mixed cultures using AMS® and conventional methods. For pure cultures, the sensitivity of AMS® ranged from 83% for *Serratia* to 99% for *E. coli.* Specificity was 98% for *E. coli* to 100% for Group D streptococci. Mixed cultures gave sensitivities of 90% (*E. coli*) to 100% (*Citrobacter*) and specificities of 98% (*S. aureus*) to 100% (*E. coli*). Challenge with organisms other than the nine most common resulted in growth of most in the nonselective medium, but misidentification of about 20% as one of the nine organisms identifiable by AMS®.[47] This appears to be an amplification of an earlier report by most of the same authors on a collaborative study involving seven laboratories.[48] Seeded cultures were tested by AMS® and conventional methods. Significant bacterial levels (≥ 7×10^4 cells per milliliter) were detected with > 90% agreement. Identification agreements ranged from 31 to 100%.

Four microplates containing different antimicrobial concentrations are available for susceptibility testing of *E. coli, Klebsiella/Enterobacter, Proteus* spp., and *P. aeruginosa.* Inocula for the susceptibility microplates may be obtained from the identification microplate. Test results are obtained after 12 hr of incubation. Overall agreement with the Kirby-Bauer method for approximately 1000 organisms was found to be ≥ 90%.[49]

D. Scattered Light Spectra

The light-scattering systems developed by Science Spectrum, Inc. (Santa Barbara, Calif.) illuminate the rewards and risks of applying high technology to microbiology. Here, a moderately dense suspension of bacteria is irradiated by a laser, while a detector traverses 180° around the suspension and measures scattered light (see Figure 2).

A few thousand cells in the cuvette contribute to the scattered light which is recorded as a function of angle θ. These scans are called differential light scattering (DLS) patterns and depend on the average size, size distribution, and structure of the cells.[50] Indeed, individual cells may be scanned while levitated by means of an electrostatic field in a vacuum.[51] Considerable information about the structure of microbes is contained in DLS patterns, although the scans are not detailed enough for species identifications. Brief exposure to a variety of media followed by rapid generation of DLS patterns may eventually provide an identification method.[52]

One practical application of the DLS principle has been found in an automated antimicrobial susceptibility system developed by Science Spectrum and called Differential III®. The dispenser module is used to load a circular tray of ten cuvettes with a Gram-positive or Gram-negative panel of antibiotics. Inoculum is prepared by incubating a suspension of colonies for 30 min and is manually added to the antimicrobial solutions. These cultures are incubated for 90 min in the Differential III® instrument before DLS patterns are automatically produced. A proprietary algorithm processes the scans and prints results in terms of sensitive, intermediate, or resistant. A total of 136 enterics encompassing 23 species were analyzed by Differential III® and Kirby-Bauer, with 96.4% agreement. A total of 36 pseudomonads gave 95% agreement.[53] The Differential III® has been licensed by Science Spectrum to a large corporation which is planning to redesign and reintroduce this instrument to the clinical market.

E. Light Scattering

Pfizer Inc. (New York) has made somewhat different use of light scattering in its Autobac 1® antimicrobial susceptibility device. It takes advantage of the fact that light scattering by particles of bacterial size peaks at approximately 30 to 40° from the incident beam (θ = 30 to 40 in Figure 2). The Autobac 1® detector is fixed at this angle. At the heart of Autobac 1® is a unique cuvette with 13 chambers. A disk loader is used to place disks containing dried antimicrobials in the cuvette and 12 different disks may be selected from a total of 24. A photometer in the instrument is used to standardize the turbidity of an inoculum made by suspending colonies in broth. Exactly 2 mℓ of inoculum is added to a tube of Eugonic broth which is attached to the cuvette and distributes the culture into the chambers containing antimicrobial disks. Cuvettes are incubated with shaking 3 to 5 hr, removed from the shaker, and read in the photometer module. Each antimicrobial sample is compared to the control, and the ratios of growth are used to interpret the results as sensitive, intermediate, or resistant. These results are printed on a card. If the control chamber has not made approximately three generations of growth, no results are printed, and the cuvette is returned to the shaker/incubator.[54]

A collaborative study involving 7 laboratories and 300 strains of bacteria compared Autobac 1® with the Bauer-Kirby disk diffusion and agar dilution methods.[55] An overall average of 91.5% agreement was obtained, with 13 drugs giving agreement of 90% or better, 3 being between 85 to 90%, and 1 (nitrofurantoin) at 77%. (Further investigation showed that the nitrofurantoin discs were out of specifications). When disagreements occurred, the Autobac 1® tended to designate as susceptible strains found to be intermediate or resistant by the other test. This was attributed to the narrow intermediate zone or to the tendency of some organisms to settle out of suspension, thus causing the instrument to underestimate growth. Another study, involving 2 laboratories and 604 isolates, compare Autobac 1® and Bauer-Kirby results.[56] Agreement ranged from 76.7 (nitrofurantoin) to 98.4% (clindamycin), with 10 of 15 antimicrobials giving greater than 90% agreement. The major discrepencies were *E. coli* with ampicillin and cephalothin and *Proteus mirabilis* with nitrofurantoin. Still another comparison of Autobac 1® and disk diffusion for 1500 Gram negatives gave agreement overall of 90%, although some strain/antimicrobial combinations had up to 30% disagreement. A total of 375 *S. aureus* isolates had 93% agreement.[57] Some strain/antimicrobial combinations with the greatest disagreements, such as *E. coli*/ampicillin[58] and *S. aureus*/methicillin[59] have been studied but without substantial reduction in the disagreements.

Ryan and Tilton[60] used Autobac 1® to determine the MICs of ampicillin for *H. influenzae*. The results for 57 isolates were compared to the MICs by agar dilution using 7 dilutions of ampicillin. A total of 65% of the isolates had exact agreement; there was agreement within one dilution in 91% of the isolates. McGowan et al.[61] used Autobac 1® to learn the "semiquantitative" antimicrobial susceptibility of 30 Gram-negative isolates using 9 antimicrobials. In this test, three antimicrobial dilutions and a control were used, giving four antimicrobial/strain relationships. This is an attempt to bridge the gap between sensitive-intermediate-resistant and complete MIC reporting. Complete agreement between Autobac 1® and agar dilution occurred in 78% of the cases. Most disagreements involved only one dilution.

Matsen et al.[62] reported an important study in which a comparison of Autobac 1® and the disk diffusion method for 300 specimens was correlated with a survey of the involved physicians. Autobac 1® and preliminary culture results were reported the afternoon of the day after receipt of specimens. The disk diffusion results were reported a day later. The survey indicated that in 32% of the Autobac 1® cases, an antibiotic other than that which would have been used otherwise was prescribed. In 12% of the cases, established therapy was changed because of the susceptibility results. In addition, physicians estimated that the patients' hospital stay was shortened in 14% of the cases. If automated microbiology is to compete successfully with other contenders for limited hospital funds, studies such as these which examine the effects of innovation on patient care will be important.

Boyd et al.[63] have examined the use of Autobac 1®-generated antimicrobial profiles to identify bacterial isolates. Pattern recognition principles were used to classify 2000 isolates on the basis of susceptibility to 9 antibiotics for Gram negatives and 11 for Gram positives. A 95% correlation overall between the susceptibility profile and conventional identification was reported. Sielaff et al.[64] used Autobac 1® to generate antimicrobial susceptibility profiles for 481 isolates of 9 Gram-negative species. A quadratic discriminate function was used to determine the identification probabilities. The use of 18 antimicrobials gave a 97% correlation with conventional methods; 14 antimicrobials allowed 95% correlation. Acquisition of resistance was cited as a potential cause of misidentification. Generation of susceptibility profiles using nonclinical inhibitors has reduced this problem.[65] A total of 33 species could be identified.[65]

F. Optical Density

The Diagnostic Division of Abbott Laboratories (Dallas, Tex.) has developed an automated antimicrobial susceptibility system, the MS-2®. It is based on detection of changes in optical density using light-emitting diodes and photo detectors. A cuvette is used which has 11 small chambers separated by a hydrophobic membrane from an upper chamber. In use the bottom chambers are loaded with antimicrobial-containing disks. A tube of medium is inoculated with several colonies and poured directly into the upper chamber. The cuvette is placed in the analysis module and incubated with shaking until logarithmic growth is detected. The culture then is dispensed automatically into the bottom chambers and incubation continues. Each chamber is monitored automatically, and the information is sent to the calculator/printer module. This module calculates the growth rate constant for each condition and from this data prints out a sensitive-resistant interpretation and a MIC (see Figure 3). Each analysis requires 3 to 5 hr. One analysis module accepts 8 cuvettes, and 7 modules may be combined

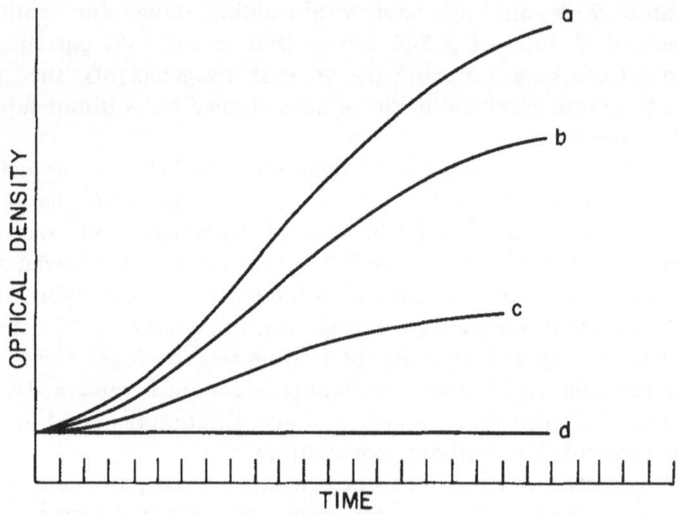

FIGURE 3. Bacterial growth monitored by MS-2®. The control module calculates the generation time and growth rate constant for each isolate. After a finite number of biological generations, the control module executes a final analysis and prints out the result for each isolate and antimicrobic combination tested. The clinically relevant MIC is determined mathematically from a linear regression analysis. (a) Control; (b) resistant (MIC ⩾3 times initial disc concentration; (c) MIC curve; (d) susceptible curve. (From MS-2® brochure, Abbott Laboratories, Inc., Dallas, Tex. With permission.)

with one calculator/printer module for a total of 56 simultaneous tests.[66] Testing with more than 200 clinical isolates has provided a correlation with agar diffusion of more than 90%.[67]

G. Plate Streaking

A machine which automatically streaks specimens on agar plates, the Autostreaker®, has been developed by Tomtec, Orange, Conn. This device stores up to 800 plates and automatically streaks liquid specimens on up to 10 different stacks. Streaking is done by a piece of sterile plastic tubing which first aspirates and then delivers sample at a preset rate while describing a pattern over a spinning plate. Evaluation of the system in terms of recovery of organisms, accuracy of counts, and colonial appearance led to favorable comparisons with hand streaking. Specimens not normally liquid were successfully extracted with Stuart's medium and streaked, as were sputum specimens after treatment with sputolysin.[68]

H. Microcalorimetry

One of the most interesting new approaches to the detection and identification of bacteria is the use of microcalorimetry. Instrumentation Laboratory Inc. (Lexington, Mass.) has developed a 40-channel instrument which monitors metabolic heat production at 37°C.[69] Approximately 10^5 to 10^6 cells per milliliter are required for generation of a heat profile which is usually characterized by an increasing rate of heat production in the log phase and declining rates thereafter. Some profiles are complete in 3 hr; others require up to 14 hr. If a complex medium is used, the profiles are characterized by peaks and valleys, presumably reflecting sequential exhaustion of nutrients. These profiles are species specific (see Figure 4). Profiles of 200 cultures including 47 species could be distinguished by eye, although *E. coli* and *Salmonella paratyphi* B unaccount-

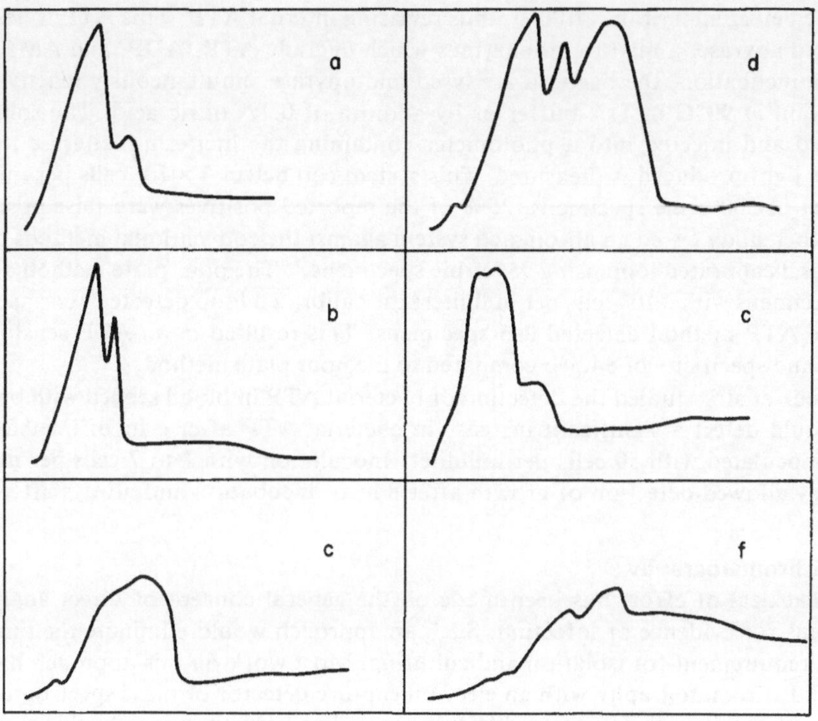

FIGURE 4. Generation of profiles by microcalorimetry. Six profiles are shown. For each, the abscissa represents time, and the ordinate represents heat production. (a) *Enterobacter aerogenes;* (b) *Klebsiella;* (c) *Proteus vulgaris;* (d) *Enterobacter cloacae;* (e) *Escherichia coli;* (f) *Proteus rettgeri.* (From Boling, E. A., Blanchard, G. C., and Russell, W. J., *Nature (London),* 241, 472, 1973. With permission.)

ably produced similar profiles. The use of another medium allowed distinction between these species.[70]

LKB (Stockholm) produces a microcalorimeter which incubates cultures from 0 to 50°C, functions aerobically or anaerobically and may be used as a flow or batch calorimeter. Bettelheim et al.[71] used a flow calorimeter to screen urine specimens for significant bacteria (> 10^5 cells per milliliter). A total of 290 samples were examined in parallel with conventional methods. No false negatives were noted, although 23% of the 190 samples giving a positive heat response contained fewer than 10^5 cells per milliliter.

I. ATP Detection

Another application of the NASA search for extraterrestrial life is the use of ATP analyses to detect the presence of microbes. The basic reaction is ATP-stimulated light production by a solution of firefly luciferin and luciferase. Photometers designed for ATP assay are produced by DuPont (Wilmington, Del.), Scientific Applications, Inc. (San Diego, Calif.), and American Instrument Company (Silver Spring, Md.). A scintillation counter often can be used to detect light production after modification. NASA has published an extensive treatment of the entire area of application of luminescence to laboratory medicine.[72]

An automated instrument which measures bacterial ATP in urine has been invented by the NASA group.[73] Urine specimens, especially those from sick people, often contain white blood cells, platelets, and red blood cells. In order to detect bacterial ATP, one must first remove all nonbacterial ATP. Human cells and platelets are ruptured with the detergent Triton X-100®, thus releasing internal ATP. This ATP is destroyed by added apyrase, a mixture of enzymes which degrade ATP, ADP, and AMP. After suitable incubation, the bacteria are lysed and apyrase simultaneously inactivated by incubation at 90°C in Tris buffer or by additon of 0.1 *N* nitric acid. The solution is decanted and injected into a photometer containing the luciferin-luciferase mixture, and the light produced is measured. This system can detect 3×10^4 cells per milliliter. In a test of 500 urine specimens, 25% of the reported positives were false positives.[74]

Another study tested an automated system against the conventional methods of pour plates and calibrated loop using 957 urine specimens.[75] The pour plate method detected 207 specimens with $>10^5$ cells per milliliter; the calibrated loop detected 164 specimens; and the ATP method detected 295 specimens. This resulted in an ATP sensitivity of 93.6% and specificity of 84.5% compared to the pour plate method.

Schrock et al.[76] studied the detection of bacterial ATP in blood seeded with bacteria. They could detect a significant increase in bacterial ATP after 6 hr of incubation of blood inoculated with 50 cells per milliliter. Inoculation with 1 to 7 cells per milliliter generally allowed detection of growth after 8 hr of incubation and always after 12 hr.

J. Gas Chromatography

A great deal of effort has been made on the general concept of direct analysis of specimens for evidence of infection. Such an approach would eliminate the time consuming requirement for isolation and culturing. Most work on this approach has used gas liquid chromatography with an electron capture detector or mass spectrograph for analysis of specimens. The most difficult aspect has been the reproducible treatment of the specimen in order to provide volatile components large enough for specificity. In addition, computer programs capable of identifying patterns in the face of changes due to column aging or replacements are necessary. Mitruka[77] studied the sera of patients and animals which were infected with a variety of bacteria. An example of gas liquid chromatograms resulting from analysis of phosphorylated sugars in sera of patients is in Figure 5. Analysis of 300 coded chromatograms from patients' sera resulted in 99% accurate identification of the infectious agent. Gas chromatographic analysis of sera of viral infected animals also shows promise. Brooks discusses the general problem in some detail and emphasizes the fact that both microbial components and metabolites, as well as host metabolic products produced in response to infection are involved in the chromatograms.[78,79] The technique may be useful for rapid identification of bacteria in cerebrospinal[80] and synovial fluids.[81] If bacteria growing in liquid media are to be identified by gas chromatography, it is important that the same lot of growth media be used throughout.

Single bacterial colonies on agar media may be sampled using glass spheres coated with tetramethylammonium hydroxide. A sphere is sealed in a capsule which is heated in the gas chromatograph at 320°C, causing derivatization and volatilization of the bacteria.[82] This technique has been used to characterize ten species of mycobacteria[83] and several *Pseudomonas* species.[84]

K. Specific Enzymatic Profiles

Another approach to automation of bacterial identification exploits the fact that many enzymatic processes may be assayed by the measurement of a common end product. Bascomb and Grantham[85,86] studied the rapid identification of *Proteus* species using a three-channel Technicon Inc. (Tarrytown, N.Y.) analyzer. The three channels were used to assay ammonium production by an untrasonically lysed bacterial suspension, ammonium from a control lysate-free solution, and total protein of the bacterial lysate. The use of fifteen L-amino acids as substrates allowed the generation of specific

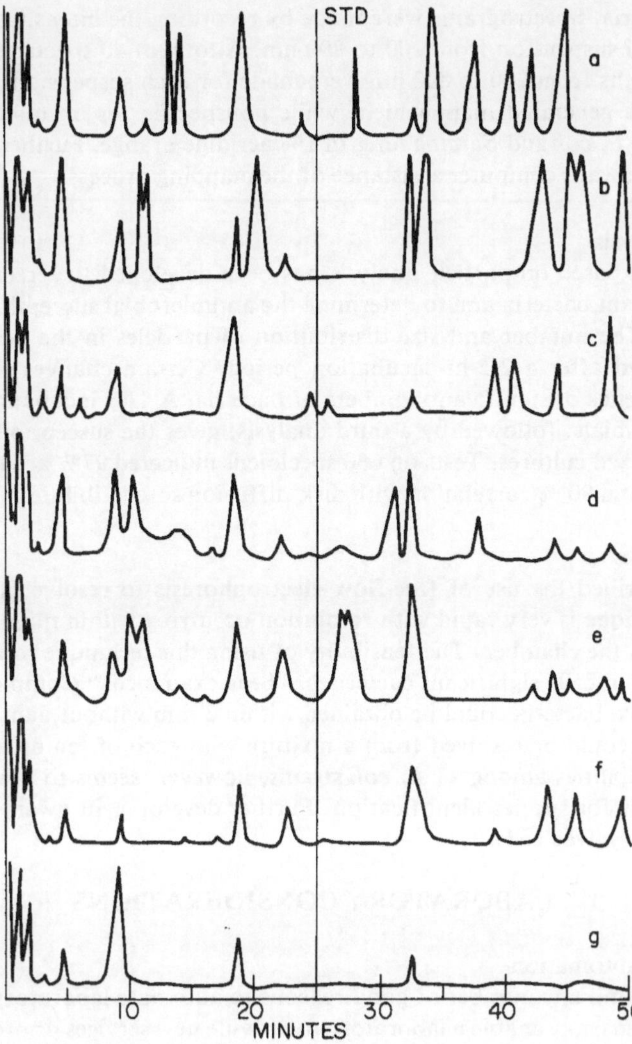

FIGURE 5. Gas chromatography patterns of sera from patients infected with bacteria. The temperature was initially at 90°C and was increased at 4°/min until reaching 230°C where it was held constant. Infected patients were as follows: (a) *Staphylococcus aureus;* (b) *Salmonella enteritidis;* (c) Group D *Streptococcus;* (d) Group A *Streptococcus;* (e) *Streptomyces viridans;* (f) Alpha-hemolytic *Streptococcus;* (g) uninfected control. (From Mitruka, B. M. *New Approaches to the Identification of Microorgansims*, Heden, C. G. and Illeni, T., Eds., Academic Press, New York, 1975. With permission.)

profiles for the six *Proteus* species tested. Reproducibility of profiles within strains of species generally was good.

Two other classes of enzymes have been included in this type of analysis — nitrophenyl releasing enzymes, including esterases, endopeptidases, galactosidases and phosphatases, and enzymes which reduce a tetrazolium salt, a variety of sugar and amino acid dehydrogenases. Two analyzer channels assayed 18 different enzymes, and the third channel determined protein content.[87]

L. Fluorescent Spectrograms

Ginnell and Feuchtbaum[88] have described the use of fluorescence spectrograms for identifying bacteria. Spectrograms were made by recording the intensity of light emitted by a bacterial suspension from 200 to 800 nm. A total of 40 spectrograms at stimulating wavelengths from 200 to 800 nm were made for each suspension. A plot of the iso-intensity lines generated maps which, while nonspecific for unstained cells, were quite specific for *E. coli* and *Sarcina lutea* in 1% acridine orange. Further development of this approach awaits computer assistance of the mapping process.

M. Particle Analysis

An instrument based on particle analysis has been developed to screen urine specimens for significant bacteria and to determine the antimicrobial susceptibilities of bacteria present.[89] The number and size distribution of particles in the urine are determined before and after a 2.5-hr incubation period. Certain changes in the pattern indicate the presence of significant numbers of bacteria. A 1-hr incubation after addition of antimicrobials, followed by a third analysis, gives the susceptibilities and any indications of mixed cultures. Tests on 650 specimens indicated 97% accuracy in determining bacteria and 90 % correlation with disk-diffusion susceptibilities.

N. Electrophoresis

Hannig[90] described the use of free-flow electrophoresis to resolve microbial mixtures. This technique is very rapid with resolution occurring within minutes of sample introduction into the chamber. The feasibility of using this technique to resolve bacterial mixtures of clinically significant bacteria has been examined.[91] Complete resolution of mixtures of two bacteria could be obtained within 2 min without damage or loss of the cells. *E. coli* could be resolved from a mixture with each of ten different species. Variations in mobilities among 23 *E. coli* strains, however, seems to preclude the use of electrophoresis for species identification. Further development awaits rapid means for detection or resolved cells.

IV. LABORATORY CONSIDERATIONS

A. Reasons for Automation

One may expound any number of good reasons to automate laboratory procedures. Automated systems may enable a laboratory to provide new services or provide services quicker, with increased reliability, in greater abundance, or at lower cost. Any of these benefits may justify consideration of an automated instrument. In any case, the total economic effect must be considered. While one may assume that every automated device on the market will reduce costs under some conditions in some laboratories, it is essential that a major investment be preceded by a careful cost analysis.

B. Financial Analysis

The most informative way to evaluate the financial effects of proposals is through some form of discounted cash flow. These methods are well developed and described in textbooks.[92] The steps in one useful method, the Net Present Value, are summarized as follows:

1. Estimate the useful lifetime of the system.
2. Estimate the net cash flow for each of these years. The net cash flow is defined as profit after taxes plus depreciation. In the case of a hospital, the net cash flow would be the difference in the laboratory profit or loss with and without the instrument plus depreciation. This figure must include consideration of supplies, labor, maintenance, and overhead. If labor savings are anticipated, it is important that the savings are realizable, i.e., that labor saved be used on projects of

equal value and not wasted. The final year cash flow will include an estimate of the terminal value of the instrument, if any.

3. Discount future cash flows at the cost of capital. This step is not as mysterious as it may seem. The cost of capital for our purposes is the interest rate (after taxes) which must be paid for a loan. The present value of each of the projected yearly cash flows is determined using the cost of capital. For example, if the cost of capital is 6%, the present value (today) of $1 (to be received) 1 year from now is $0.943. That means that $0.943 invested at 6% interest would yield $1 in 1 year.
4. Determine the Net Present Value. The present values of the cash flows for all the years are summed. This is the present value of the entire project. From this figure, one subtracts the cost of the equipment if it were bought and paid for today. The net present value represents the financial effect of the entire project. If positive, the project will make money; if negative, it will cost money. Decisions on automated microbiology projects will never be based solely on financial considerations. Nevertheless one should have a firm understanding of the effect of projects on costs. In addition, when capital is limited, comparison of the Net Present Values of competing projects can aid in making decisions between them. Leasing equipment rather than purchasing may be treated in a similar fashion — it is easier to estimate costs when leasing because uncertainty concerning terminal value is eliminated. There may be some tax advantages in leasing, but these are irrelevant to nonprofit institutions.

It is a reasonable extension of the discounted cash flow concept to include consideration of benefits to the patient in terms of decreased hospital stay and decreased morbidity and mortality. The key here is making realistic monetary valuations of these benefits. One may expect this type of evaluation to develop as institutions with limited capital are forced to choose between competing projects.

C. Purchase Considerations

The ability of an automated system to meet performance specifications is the most important consideration. This is hard to assess, but, fortunately, an informal group of clinical microbiologists organized through the Center for Disease Control have undertaken the collaborative evaluation of new instruments. Their findings are reported in the scientific literature[55] and provide independent evaluation of systems in a variety of clinical environments.

Another consideration is the inherent reliability of the instrument and the degree of required maintenance. The manufacturer's capacities for major maintenance, repair of malfunctions, and for development of new applications for the instrument are also important.

Many automated microbiology devices incorporate unique consumable containers or reagents. It is of utmost importance that customers be assured a reliable supply of these consumables. There is a tendency for customers to use consumables purchased through the instrument manufacturer, despite a lack of objective differences in performance between these items and those from other sources. This "captive market" allows high margins on consumables for instrument manufacturers and discourages entry of competitors. One effect is to make the customer dependent on a single source for consumables.

D. Federal Regulation

The role of the Food and Drug Administration (FDA) in automated microbiology is described by the Medical Device Amendment of 1976. It provides that all manufacturers producing medical devices after May 28, 1976 must provide the FDA with copies of labels and descriptive materials concerning the device and the manufacturer's opinion on the class under which the device should be listed. The classes are

1. Commonly used devices, such as, tongue depressors or thermometers
2. Standard devices; devices for which performance specifications will be written (No standards have as yet been completed.)
3. Premarket clearance items, generally life-sustaining or supporting devices such as pacemakers. (If the FDA agrees with the classification, the item may be marketed. It is expected that instruments for automated microbiology will fall into Class 2, although systems using novel technology may be placed in Class 3.)

V. FUTURE PROJECTIONS

The only certain projection is that this field is growing rapidly and many new devices and approaches may be expected. One may anticipate continuation of a pattern in which invention is done in university and government institutions, while development and commercialization is done by large established manufacturers. It is particularly difficult for a small company to do the nonmanufacturing functions, such as development, service, distribution, and marketing. We have seen the beginning of a shift in marketing emphasis. In the past, market development was directed toward the needs and desires of the laboratory director and staff. In the future, there will be a greater emphasis on the needs of the "end-user", the physician who uses the data.

It is curious to note that most of the automated devices on the market or undergoing development do not depend on "exotic" methods of measurement and differentiation. Rather, they are based on "classical" methods, such as measuring growth by turbidity and differentiating between species by selective growth media. These methods are automated by sophisticated computer control and analysis. It may be that a "second generation" of automated devices based on nongrowth techniques will emerge, but this is not in the immediate future.

REFERENCES

1. **Palmer, W. J. and LeQuesne, S. E.,** *Rapid Methods and Automation in Microbiology and Immunology: A Bibliography,* Information Retrival, London, 1976, unpaged.
2. **Heden, C. G. and Illeni, T., Eds.,** *New Approaches to the Identification of Microorganisms,* J. Wiley & Sons, New York, 1975, 466.
3. **Heden, C. G. and Illeni, T., Eds.,** *Automation in Microbiology and Immunology,* J. Wiley & Sons, New York 1975.
4. **Johnston, H. H. and Newsom, S. W. B.,** *Rapid Methods and Automation in Microbiology,* Learned Information, Oxford, 1976, 318.
5. **Balows, A., Ed.,** *Current Techniques for Antibiotic Susceptibility Testing,* Charles C Thomas, Springfield, Ill., 1975, 173.
6. **Schlessinger, D., Ed.,** *Microbiology — 1975,* American Society for Microbiology, Washington, D.C., 1975, 521.
7. **Isenberg, H. D. and MacLowry, J. D.,** Automated methods and data handling in bacteriology, *Annu. Rev. Microbiol.,* 30, 483, 1976.
8. **Facklam, R. R.,** Streptococci, in *Manual of Clinical Microbiology,* 2nd ed., Lennette, E. H., Spaulding, E. H., and Truant, J. P., Eds., American Society for Microbiology, Washington, D.C., 1974, 96.
9. **Galen, R. S. and Gambino, S. R.,** *Beyond Normality; the Predictive Value and Efficiency of Medical Diagnoses,* J. Wiley & Sons, New York, 1975, 82.

10. **Bartlett, R. C.,** *Medical Microbiology. Quality Cost and Clinical Relevance,* J. Wiley & Sons, New York, 1974, 9.
11. **Edwards, P. R. and Ewing, W. H.,** *Identification of Enterobacteriaceae,* 3rd ed., Burgess, Minneapolis, 1972, 67, 324.
12. **Neter, J., Wasserman, W., and Whitmore, G. A.,** *Fundamental Statistics,* 4th ed., Allyn and Bacon, Boston, 1973, 443.
13. **Lapage, S. P., Bascomb, S., Willcox, W. R., and Curtis, M. A.,** Identification of bacteria by computer: general aspects and perspectives, *J. Gen. Microbiol.,* 77, 273, 1973.
14. **Bascomb, S., Lapage, S. P., Curtis M. A., and Willcox, W. R.,** Identification of bacteria by computer: identification of reference strains, *J. Gen. Microbiol.,* 77, 291, 1973.
15. **Willcox, W. R., Lapage, S. P., Bascomb, S., and Curtis, M. A.,** Identification of bacteria by computer: theory and programming, *J. Gen. Microbiol.,* 77, 317, 1973.
16. **Dybowski, W. and Franklin, D. A.,** Conditional probability and the identification of bacteria. A pilot study, *J. Gen. Microbiol.,* 54, 215, 1968.
17. **Gyllenberg, H. G. and Niemela, T. K.,** Basic principles in computer-assisted identification of microorganisms, in *New Approaches to the Identification of Microorganisms,* Heden, C. G. and Illeni, T., Eds., J. Wiley & Sons, New York, 1975, 201.
18. **Thrupp, L. D.,** Significance and interpretation of quantitative antimicrobial susceptibility testing in clinical microbiology, in, *Current Techniques for Antibiotic Susceptibility Testing,* Balows, A., Ed., Charles C Thomas, Springfield, Ill., 1974, 94.
19. **Thornsberry, C.,** The agar diffusion antimicrobial susceptibility test, in *Current Techniques for Antibiotic Susceptibility Testing,* Balows, A., Ed., Charles C Thomas, Springfield, Ill., 1974, 6.
20. **Ericsson, H. M. and Sherris, J. C.,** Antibiotic sensitivity testing: report of an international collaborative study, *Acta Pathol. Microbiol. Scand. Suppl.,* 217, 1, 1971.
21. **Balows, A.,** The need for automation in clinical microbiology with special emphasis on antibiotic susceptibility testing, in *Advances in Automated Analysis,* Vol. 1, Thurman Associates, Miami, 1971, 367.
22. **Gavan T. L. and Butler, D. A.,** An automated microdilution method for antimicrobial susceptibility testing, in *Current Techniques for Antibiotic Susceptibility Testing,* Balows, A., Ed., Charles C Thomas, Springfield, Ill., 1974; 88.
23. **Levin, G. V. and Straat, P. A.,** Viking labeled release biology experiment: interim results, *Science,* 194, 1322, 1976.
24. **DeLand, F. H. and Wagner, H. N.,** Early detection of bacterial growth with carbon-14 labelled glucose, *Radiology,* 92, 154, 1969.
25. **DeLand, F. H. and Wagner, H. N.,** Automated radiometric detection of bacterial growth in blood cultures, *J. Lab. Clin. Med.,* 75, 529, 1970.
26. **Coleman, R. M., Laslie, W. W., and Lambe, D. W.,** Clinical comparison of aerobic, hypertonic, and anaerobic culture media for the radiometric detection of bacteremia, *J. Clin. Microbiol.,* 3, 281, 1976.
27. **Wood, N. G.,** Comparative study of two systems for detection of bacteraemia and septicaema, *Br. J. Clin. Pathol.,* 29, 530, 1976.
28. **Randall, E. L.,** Long-term evaluation of a system for radiometric detection of bacteremia, in *Microbiology - 1975,* Schlessinger, D., Ed., American Society for Microbiology, Washington D.C., 1975, 39.
29. **Schrot, J. R., Hess, W. C., and Levin, G. V.,** A method for the rapid radiorespirometric identification of bacteria, *Proc. Am. Soc. Microbiol.,* M280, 113, 1974.
30. **Levin, G. V. and Takeguchi, M. M.,** personal communication.
31. **Bourgeois, C. and Mafart, P.,** The uptake of a labelled metabolite as a technique for microorganisms detection, in *Rapid Methods and Automation in Microbiology,* Johnston, H. H and Newsom, S. W. B., Eds., Learned Information, Oxford, 1976, 4.
32. **Stewart, G. N.,** The changes produced by the growth of bacteria in the molecular concentration and electrical conductivity of culture media, *J. Exp. Med.,* 4, 235, 1899.
33. **Ur, A. and Brown, D. F. J.,** Monitoring of bacterial activity by impedance measurements, in *New Approaches to the Identification of Microorganisms,* Heden, C. G. and Illeni, T., Eds., J. Wiley & Sons, New York, 1975, 61.
34. **Cady, P.,** Rapid automated bacterial identification by impedance measurements, in *New Approaches to the Identification of Microorganisms,* Heden, C. G. and Illeni, T., Eds., J. Wiley & Sons, New York, 1975, 73.
35. **Hadley, W. K. and Senyk, G.,** Early detection of microbial metabolism and growth by measurement of electrical impedance, in *Microbiology — 1975,* Schlessinger, D., Ed., American Society for Microbiology, Washington, D. C., 1975, 12.

36. **Ur, A.**, Prospects and limitation of the impedence technique, in *Rapid Methods and Automation in Microbiology*, Johnston, H. H. and Newsom, S. W. B., Eds., Learned Information, Oxford, 1976, 245.
37. **Nunke, B., Dufour, S. W., and Kraeger, S. J.**, Bacterial enumeration in urine specimens by automated impedance measurements, *Proc. Am. Soc. Microbiol.*, C25, 31, 1975.
38. **Specter, S., Throm, R., and Friedman, H.**, Screening for bacteriuria in the clinical microbiology laboratory by electrical/impedance monitoring, in *Rapid Methods and Automation in Microbiology*, Johnston, H. H. and Newsom, S. W. B., Eds., Learned Information, Oxford, 1976, 13.
39. **Hadley, W. K.**, The use of electrical impedance measurements for detection of microorganisms in blood cultures, in *Rapid Methods and Automation in Microbiology*, Johnston, H. H. and Newsom, S. W. B., Eds., Learned Information, Oxford, 1976, 14.
40. **Kahn, W., Friedman, G., Rodriguez, W. Contronі, G., and Ross, S.**, Rapid detection and isolation of bacteria in blood and cerebrospinal fluids in children by the electrical impedance method, in *Rapid Methods and Automation in Microbiology*, Johnston, H. H. and Newsom, S. W. B., Eds., Learned Information, Oxford, 1976, 14.
41. **Hardy, D. L., Cox, M. E., and Lindberg, L. H.**, Automated monitoring of anaerobic bacteria by impedance measurements, *Proc. Am. Soc. Microbiol.*, C98, 42, 1976.
42. **Wheeler, T. G. and Goldschmidt, M. C.**, Determination of bacterial cell concentrations by electrical measurements, *J. Clin. Microbiol.*, 1, 25, 1975.
43. **Goldschmidt, M. C. and Wheeler, T. G.**, Determination of bacterial cell concentration in urine specimens by electrical measurements, in *Microbiology — 1975*, Schlessinger, D., Ed., American Society for Microbiology, Washington, D.C., 1975, 6.
44. **Jones, P. W., Aldridge, C., Charles, R. A., and Frenkel, J. B.** An automated microbiological detection/identification system. I. System description, *Proc. Am. Soc. Microbiol*, C73, 38, 1976.
45. **Aldridge, C., Gibson, S., Lanham, J., Meyer, M., Vannest, R., Jones, P., and Chalres, R.**, An automated microbiological detection/identification system. II. Microbiological considerations, *Proc. Am. Soc. Microbiol.*, C74, 38, 1976.
46. **Sonnenwirth, A. C.**, Developmental studies of an automated microbial detection and identification system, *Proc. Am. Soc. Microbiol.*, C75, 38, 1976.
47. **Smith, P. B., Balows, A., Washington, J., Isenberg, H., Gavin, T., Sonnenwirth, A., and Taylor, W.**, A collaborative evaluation of the automicrobic system, *Proc. Am. Soc. Microbiol.*, C102, 52, 1977.
48. **Smith, P. B., Gavan, T. L., Sonnenwirth, A., Isenberg, H. D., Taylor, W. I., and Washington, J. A.**, The automated microbial system — bacteriological evaluation, in *Rapid Methods and Automation in Microbiology*, Johnston, H. H. and Newsom, S. W. B., Eds., Learned Information, Oxford, 1976, 37.
49. **Fadler, N., Jones, P. W., and Gibson, S.**, Developmental study of automated susceptibility testing using the Automicrobic System®, *Proc. Am. Soc. Microbiol.*, C32, 40, 1977.
50. **Anon.**, *Differential light scattering. Antibiotic susceptibility testing and the Differential III System*, Science Spectrum, 1974, 1.
51. **Wyatt, P. J. and Phillips, D. T.**, Structure of single bacteria from light scattering, *J. Theoret. Biol.*, 37, 493, 1972.
52. **Wyatt, P. J.**, The detection and identification of bacteria by laser light scattering, in press.
53. **Coe, A. E. and Kahn, M. R.**, Automated versus conventional susceptibility testing: a parallel trial in the clinical laboratory, *Proc. Am. Soc. Microbiol.*, C32, 32, 1975.
54. **Anon.**, Autobac 1® package insert, Pfizer Diagnostic Division, Pfizer Inc., New York, 1977, 1.
55. **Thornsberry, C., Gavan, T. L., Sherris, J. C., Balows, A., Matsen, J. M., Sabath, L. D., Schoenknecht, F., Thrupp, L. D., and Washington, J. A.**, Laboratory evaluation of a rapid, automated susceptibility testing system: report of a collaborative study, *Antimicrob. Agents Chemother.*, 7, 466, 1975.
56. **Phillips, S. B., Greenberg, D. S., Sierra, M. F., and Amsterdam, D.**, Autobac 1/Bauer-Kirby comparison and reproducibility: an interhospital evaluation of 604 clinical isolates, *Proc. Am. Soc. Microbiol.*, C31, 40, 1977.
57. **Stubbs, K. G. and Wicher, K.**, Laboratory evaluation of the automated antimicrobial susceptibility system — Autobac 1, *Proc. Am. Soc. Microbiol.*, C110, 44, 1976.
58. **Greenberg, R. L., Gottschall, R. L., and Randall, E. L.**, A comparison of Autobac 1 results with early and conventional readings of Bauer-Kirby disc diffusion tests, *Proc. Am. Soc. Microbiol.*, C111, 1975.
59. **Cleary, T. J.**, Methicillin-resistant *Staphylococcus aureus* susceptibility testing using an automated system, Autobac 1, *Proc. Am. Soc. Microbiol.*, A78, 14, 1977.
60. **Ryan, R. and Tilton, R. C.**, Rapid method for determining the minimum inhibitory concentration of ampicillin for *Haemophilus influenzae*, *Antimicrob. Agents Chemother.*, 11, 114, 1977.

61. **McGowan, J. E., Ayres, C., and Sabath, L. D.,** Rapid semiquantitative testing of antibiotic susceptibility: use of a multicell disc elution system, *Antimicrob. Agents Chemother.*, 7, 543, 1975.
62. **Matsen, J. M., Krall, B. J., and Saxon, B. A.,** The influence of first day susceptibility testing results on antimicrobial therapy and length of patient hospital stay, in *Rapid Methods and Automation in Microbiology*, Johnston, H. H. and Newsom, S. W. B., Eds., Learned Information, Oxford, 1976, 60.
63. **Boyd, J. C., Lewis, J., Man, J. J., Harper, A. M., and Kowalski, B. R.,** Identification of microorganisms by pattern recognition of antibiotic sensitivity profiles, in *Rapid Methods and Automation in Microbiology*, Johnston, H. H. and Newsom, S. W. B., Eds., Learned Information, Oxford, 1976, 55.
64. **Sielaff, B. H., Johnson, E. A., and Matsen, J. M.,** Computer assisted bacterial identification utilizing antimicrobial susceptibility profiles generated by Autobac 1, *J. Clin. Microbiol.*, 3, 105, 1976.
65. **Sielaff, B. H., Buck, G. E., and Matsen, J. M.,** An innovative approach to bacterial identification using Autobac 1, *Proc. Am. Soc. Microbiol.*, C-101, 52, 1977.
66. **Stockert, J. E., Wilborn, R. C., and Fekete, N. M. G.,** Rapid automated anitmicrobial susceptibility testing, Test instrument description (MS-2™), *Proc. Am. Soc. Microbiol.*, C128, 56, 1977.
67. **Spencer, H. J., Welaj, P. M., and Swensen, L. M.,** Rapid automated antimicrobial susceptibility testing. Comparison of results from an automated instrument system (MS-2™) with standard methods, *Proc. Am. Soc. Microbiol.*, C129, 57, 1977.
68. **Tilton, R. C. and Ryan, R. W.,** Automated primary plating in clinical microbiology. An evaluation report, *Proc. Am. Soc. Microbiol.*, C99, 52, 1977.
69. **Russell, W. J., Zettler, J. F., Blanchard, G. C., and Boling, E. A.,** Bacterial identification by microcalorimetry, in *New Approaches to the Identification of Microorganisms*, Heden, G. C. and Illeni, T., Eds., J. Wiley & Sons, New York, 1975, 101.
70. **Boling, E. A., Blanchard, G. C., and Russell, W. J.,** Bacterial identification by microcalorimetry, *Nature (London)*, 241, 472, 1973.
71. **Bettelheim, K. A., O'Farrell, S. M., Al-Salihi, S., Shaw, E. J., and Beezer, A. E.,** Clinical investigation into the potential use of microcalorimetry for the diagnosis of bacteriuria, in *Rapid Methods and Automation in Microbiology*, Johnston, H. H. and Newson, S. W. B., Eds., Learned Information, Oxford, 1976, 8.
72. **Picciolo, G. L., Chappelle, E. W., Deming, J. W., McGarry, M. A., Nibley, D. A., Okrend, H., and Thomas, R. R.,** Applications of luminescent systems to infectious disease methodology. X-726-76-212. Goddard Space Flight Center, Greenbelt, Md., 1976, 1.
73. **Chappelle, E. W. and Picciolo, G. L.,** Method of Detection and Counting Bacteria in Body Fluids, U.S. patent 3,745,090, 1973.
74. **Gutekunst R. R., Jaffee, B., and Chappelle, E. W.,** Detection of bacteria in urine by the firefly luciferase ATP assay, *Proc. Am. Soc. Microb.*, C72, 38, 1976.
75. **Johnston, H. H., Mitchell, C. J., and Curtis, G. D. W.,** An automated test for the detection of significant bacteriuria, *Lancet*, 7981, 400, 1976.
76. **Schrock, C. G., Barza, M. J., Deming, J. W., Picciolo, G. L., Chappelle, E. W., and Weinstein, L.,** Rapid detection of bacterial growth in blood cultures by means of adenosine triphosphate determination, *Proc. Am. Soc. Microbiol.*, C71, 37, 1976.
77. **Mitruka, B. M.,** Rapid automated identification of microorganism in clinical specimens by gas chromatography, in *New Approaches to the Identification of Microorganisms*, Heden, C. G. and Illeni, T., Eds., John Wiley & Sons, New York, 1975, 123.
78. **Brooks, J. B.,** Detection of bacterial metabolites in spent culture media and body fluids by electron capture gas-liquid chromatography, *Adv. Chromatogr.*, in press.
79. **Brooks, J. B.,** Identification of disease and disease-causing agents by analysis of spent culture media and body fluids with electron capture gas-liquid chromatography, in *Microbiology — 1975*, Schlessinger, D., Ed., American Society for Microbiology, Washington D.C., 1975, 45.
80. **Schlossberg, D., Brooks, J. B. and Schulman, J. A.,** Possibility of diagnosing meningitis by gas chromatography; cryptococcal meningitis., *J. Clin. Mecrobiol.*, 3, 239, 1976.
81. **Brooks, J. B., Kellogg, D. S., Alley, C. C., Short, H. B., Handsfield, H. H., and Huff, B.,** Gas chromatography as a potential means of diagnosing arthritis. I. Differentiation between staphylococcal, streptococcal, gonococcal and traumatic arthritis, *J. Infect. Dis.*, 129, 660, 1974.
82. **Wade T. J., Mandle, R. J., and Schaedler, R. W.,** Further automation of the gas-liquid chromatographic characterization of tetramethylammonium hydroxide treated bacteria, *Proc. Am. Soc. Microbiol.* M282, 113, 1974.
83. **Ohashi, D. K., Wade, T. J., and Mandle, R. J.,** Characterization of ten species of mycobacteria by gas chromatography using tetramethylammonium hydroxide, *Proc. Am. Soc. Microbiol.*, M242, 106, 1974.

84. **Wade, T. J. and Mandle, R. J.**, New gas chromatographic characterization procedure: preliminary studies on some *Pseudomonas* species, *Appl. Microbiol.*, 27, 303, 1974.
85. **Bascomb, S. and Grantham, C.**, Application of automated assay of asparaginase and other ammonia-releasing enzymes to the identification of bacteria, in *Some Methods for Microbiological Assay*, Board, R. C. and Lovelock, D. W., Eds., Academic Press, New York, 1975, 29.
86. **Bascomb, S. and Grantham, C.**, Specific enzyme profile, an automated method for bacterial classification, Abstracts First International Congress for Bacteriology, Jerusalem, Vol. 2, 1973, 132.
87. **Bascomb, S.**, Rapid identification of bacteria from clinical specimens by continuous flow analysis, in *Rapid Methods and Automation in Microbiology*, Johnston, H. H. and Newsom, S. W. B., Eds., Learned Information, Oxford, 1976, 53.
88. **Ginell, R. and Feuchtbaum, R. J.**, Fluorescent spectrophotometry in the identification of bacteria, *J. Appl. Bacteriol.*, 35, 29, 1972.
89. **Gall, L. S. and Curby, W. A.**, Instrumented approach to microbiological analysis of urine specimens, *Proc. Am. Soc. Microbiol.*, C88, 40, 1976.
90. **Hannig, K.**, Free-flow electrophoresis, in *Methods in Microbiology*, Vol. 5B, Norris, J. R. and Ribbons, D. W., Eds., Academic Press, New York, 1971, 513.
91. **Ramsey, W. S.**, Resolution of microbial mixtures by free flow electrophoresis, American Chemical Society Annual Meeting, M-059, 1976.
92. **Weston, J. F. and Brigham, E. F.**, *Managerial Finance*, 5th ed., Dryden Press, Hinsdale, Ill., 1975, 915.

AUTOMATION OF RADIOIMMUNOASSAY

Charles H. Rogers and Laughton E. M. Miles

TABLE OF CONTENTS

I. Introduction 128

II. Approaches to Automation 128

III. Filtration Techniques 129

IV. Separation by Charge or Molecular Weight 133

V. Solid Phase Antibody in Columns 136

VI. Antibody Immobilized on Magnetic Particles 138

VII. Antibody Coated on Plastic Tubes 140

VIII. Homogeneous Assays 142

IX. Conclusions 143

References 145

I. INTRODUCTION

This chapter reviews automation of radioimmunoassay (RIA) procedures in the clinical laboratory. Automation of RIA procedures is in a state of flux at this time. Only 15 years ago, competitive protein binding assays were the only method available for assaying substances in picogram per milliliter quantities. In the early 1970s the radioimmunoassay procedure using iodine-125-labeled derivatives was able to improve the assay by combining the specific characteristics of antibodies with the sensitivity of the higher specific activity, iodine-125. As procedures which had been developed in the research laboratories moved into the clinical lab, manufacturers of kit materials made available reagents for performing a wide variety of radioisotope assays. As demand for these assays increased, so did the requirements to have highly skilled technical people to perform the work. Technologists had to understand the intricacies of the various protocols, handle radioactive materials in repetitive dispensing operations, and perform data analysis by relating the unknowns to standard curves. Today some of the assays for thyroid function, cortisol, or for digoxin are in such high demand that they approach the levels of request of some of the common assays of the clinical chemistry laboratory which have been automated for several years. The promise of faster throughput, lower cost per test, better precision, and a reduction in the skill level and training required to perform the assay brings automation of RIA procedures into the picture.

Since the original paper by Berson and Yalow, there have been numerous publications describing RIA techniques and improvements in the methodology.[1-4] We will discuss only those technologies which lend themselves to the automation of RIA procedures. We will then describe instrumentation approaches which automate or highly mechanize these procedures with emphasis on commercially available devices.

II. APPROACHES TO AUTOMATION

In recent years, a substantial industry has come into being to provide complete assay kits and assay components, such as standards or trace-labeled antigens and antiserum for performing most analytical procedures. Radioactivity counters have become increasingly sophisticated in improving the rate of sample processing and the type of data reduction available. In addition, a number of automated sampling and pipetting devices are available to improve the throughput and precision of handling the reagents for these assays. Devices such as the LKB 2071, the Micromedic Systems 24000, the Packard Prias®, and the Searle 1600, have responded to the laboratories' need for mechanization and flexibility. Many laboratories have put together highly "automated" systems comprised of semiautomated or automated sample and reagent handling equipment, large capacity centrifuges, and radioactivity counters with automatic sample feeds holding 200 or more samples. These approaches will not be reviewed here. Instead, we will concentrate on those devices which provide automation of the complete process of RIA from original sample to final result.

In reviewing the steps required in automating RIA, one of the most challenging is that of separation. In the classical technique, there are many different methods for separating the "bound" from "free". It is interesting to note that the primary method of separation, that of centrifugation, has not been approached by any manufacturer of a fully automated device. Fortunately, numerous other methods for separation have been developed. Each of these methods has its own pros and cons and provides its own limitations for the designer of the equipment as well as the user. Since the separation method has significant influence for so many aspects of the procedure, we will use it as a means of categorizing the various devices available.

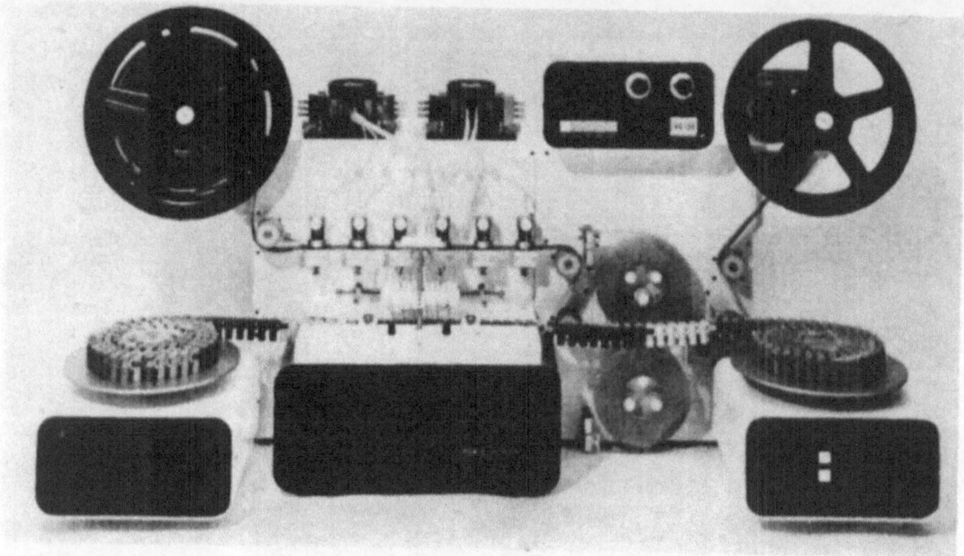

FIGURE 1. The separate components of the RIA-E 6000 system from Berthold Laboratories. (A) Reagent handling system for mixing initial reagents and for filtration and (B) BF 6000 system for counting the separate filters three at a time.

III. FILTRATION TECHNIQUES

In many commercially available assays, a precipitating reagent is added after incubation. In the so-called double-antibody method, this would include an antibody from a different species of animal which precipitates the original antibody-antigen complex by forming large networks of molecules. Other reagents which will precipitate the bound fraction include ammonium sulfate, sodium sulfate and ether, sodium sulfite, sodium chloride and/or alcohol, or polymeric substances such as polyethylene glycol. These methods tend to be fast, give excellent separation, usually 95% or better, and the reagents tend to be inexpensive. These procedures, however, can trap some of the free material in the precipitate and therefore are not as good as other methods in quantitating nanogram and picogram levels of material.

The most obvious replacement for the centrifuge in such precipitated assays is the use of a filter. Cellulose acetate, polycarbonate, glass fiber, or paper filter discs capable of entrapping submicron particles have been used in at least two approaches to automating RIA.

A system which uses the filtration technique, developed by Scriba[5] in Munich and sold by Berthold Laboratories in Wildbad, Germany, is currently available through BF Vertiebs, GmbH. Called the RIA-E 6000 system, it consists of four separate modules (see Figure 1). The pipetting operations are performed by conventional apparatus, and incubation takes place in individual vials aligned in chains of up to 650 samples. After incubation, the vials are moved to the filtration station, where six samples at a time are pumped through the filter material followed by one or two washes of the vial. Nucleopore filters of 20-mm diameter and 0.2-μm pore size were found to be superior to all other filters tested. The filters are mounted on 35-mm triacetate film in spools 1000 ft long. The film strip is sealed on both sides with adhesive plastic foil to prevent radioactive contamination (see Figure 2). After moving past the filtering station at a rate of approximately 220 samples per hour, the film strip is stored on a reel. The obvious difficulty of having to quantitate the radioactivity at this same very high rate is alleviated by providing a separate gamma counter device. The counter accepts the

FIGURE 1B.

reel of film and quantitates the filter membranes three at a time by sandwiching them between three pairs of detectors. On a number of assays, the device has provided identical standard curves to cellulose acetate filtration or centrifugation and conventional gamma counting, but provides improved precision of the replicates.

A similar system described by Bagshawe[6] at Charing Cross Hospital, London, England is currently under development by the Picker Corporation of Northford, Conn. (see Figure 3). The multiple units which comprised the system as originally described have been nicely integrated by both hardware and software means. For example, all

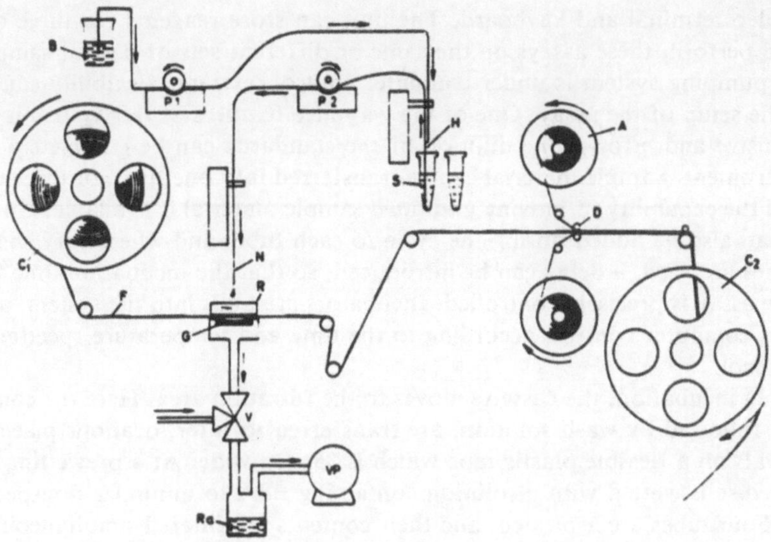

FIGURE 2.. Diagram of one filtration area as used on the RIA-E 6000, showing the film spools with individual filters, F, being placed over the suction area, R. Wash solution is delivered to the test tubes by pump P1 while all solutions are removed to the filters by P2.

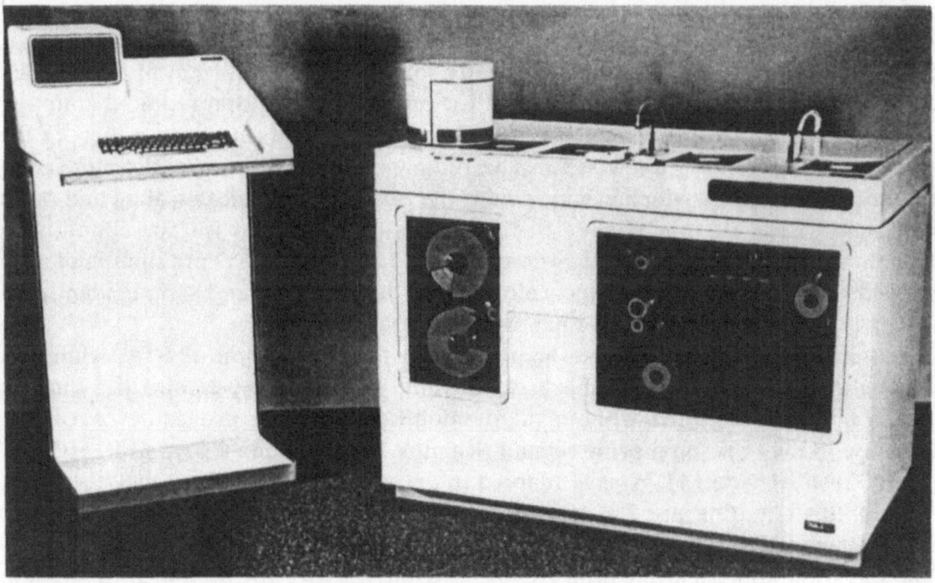

FIGURE 3. The automated RIA system based on filtration separation under development by Picker Instruments called Pace-4.

of the operations from sample dilution through incubation to final read-out are housed in one unit. Test tubes are held in convenient cassettes holding 20 racks of 8 tubes each. Each cassette can hold a complete assay, standards, controls, and samples. Excess samples can be loaded into a second cassette which the computer will identify with the standard curve determined in the first cassette. Preprogrammed assay protocols are held in the computer memory or can be entered by the operator. All control is from a video terminal and keyboard. The unit can store reagents for three differing assays and perform these assays on the same or different sets of patient samples. Because the pumping system is under computer control, extreme flexibility can be provided in the setup of the assay. One of the very nice features of this system is that the labor-intensive and error-prone dilution of the standards can be completely handled by the instrument. Sample material can be transferred into one, two, or three replicate tubes, and the capability of moving undiluted sample material is available. Up to three reagents can also be added during the cycle to each tube, and where very short incubation times are used, a delay can be introduced, so that the incubation time for each tube in the assay is precisely controlled. Incubators are built into the system, and handled under computer control, according to the time and temperature specified in the assay protocol.

Following incubation, the cassette moves to the filtration area. Here the contents of the tubes, followed by wash solution, are transferred to filter locations placed at 65-mm intervals on a flexible plastic tape which is 35 mm wide. At a prewetting station, each filter disc is wetted with a solution containing BSA to minimize nonspecific absorption. Four tubes are aspirated, and their contents are filtered simultaneously, and filtration of the contents of the 4 tubes takes 30 to 60 sec, depending on sample volume and the number of washes. Thus, the average time per tube is from 8 to 15 sec. After filtration, the tape moves through a drying chamber; the filter side is covered by an adhesive transparent tape. The four-filter strip then enters the counting unit. As in the Berthold unit, the tape filters are counted by pairs of scintillation detectors mounted above and below the tape. To minimize electrical noise and nuclear background, each set of four photomultiplier tubes (above and below the tape) with their associated crystals, shielding, and preamplifiers are built as a single assembly. The problem of keeping all eight detectors in balance is neatly solved by the stratagem of letting each filter be counted by each pair of detectors for one fourth the total required counting time. Minor imbalances between pairs are thus negated, as all filters are subject to the same, total nuclear detection system. The radioactivity counters are also under the control of the computer which even provides the function of nuclear scaling and pulse height analysis for all four channels. This makes possible dual isotope counting to handle the newer combination tests where iodine-125 and cobalt-57 are simultaneously employed. The original input sequence of standards, controls, and sample identification data is used in developing the final printout of results.

Filtration systems clearly offer a high potential for automation of RIA, while still providing the user with considerable assay flexibility. Almost any competitive binding RIA can be accommodated without major modification in its procedure. Assays in routine use in the U.K. on a prior version (Kemtek 3000) include CEA, AFP, HCG-β, TSH, HGH, FSH, and LH. Assays adapted to the system from commercially available kits as of this counting are T-4 (Corning), digoxin (Kallestad), free T-4 (Corning), cortisol (Diagnostic Products Corp.), and insulin (Corning).

The filtration approach has been found most satisfactory for double antibody and glass bead solid phase assays, with some difficulties encountered with charcoal separation techniques. Only a few types of assays would be difficult to implement using the filtration approach. For example, a double antibody method which precipitates the primary antibody prior to use with samples or a two-site IRMA in which the second reaction occurs after separation would not appear possible using this equipment. Nevertheless, the filtration technique, in conjunction with sophisticated sample handling and radioactivity counting equipment, appears to be the most versatile approach to automating RIA.

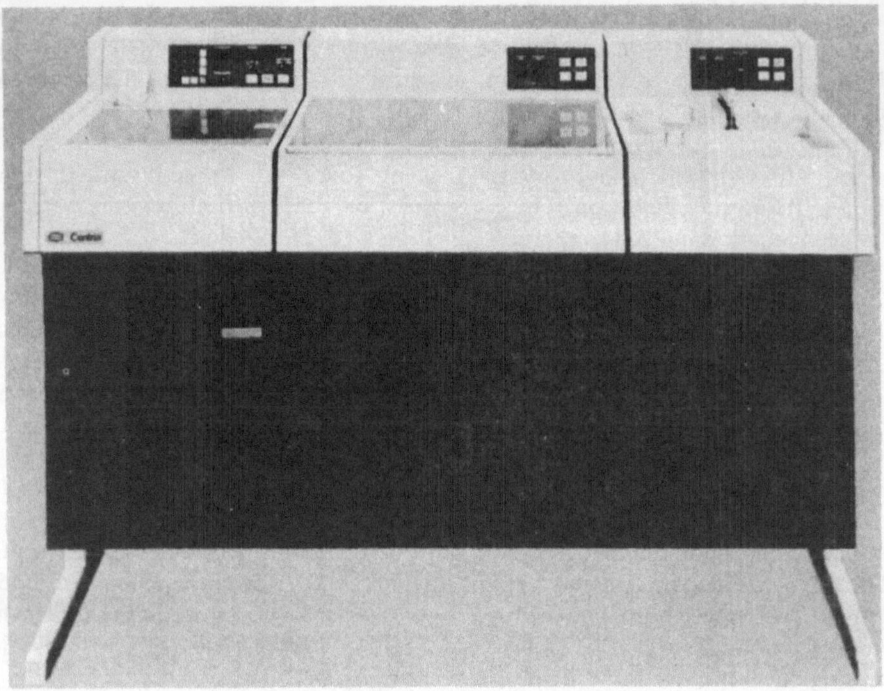

FIGURE 4. Centria® developed by Union Carbide. The reagent pipetting is done in the module on the right. Incubation and separation are in the center module, while counting is done in the module shown on the left.

IV. SEPARATION BY CHARGE OR MOLECULAR WEIGHT

The characteristics of the bound antigen-antibody complex are considerably different from the free antigen. The complex is only weakly charged, and the molecular weight is usually much larger than that of the antigen alone. This, of course, is not always the case and sometimes limits the application of this technique. For example, if the antigen is a high molecular weight protein, then binding it with a high molecular weight antibody does not significantly alter its size or its charge. Nevertheless, a considerable fraction of the work in the RIA laboratory involves assays of haptens which are well suited to separation based on charge or molecular weight.

The first commercially available automated RIA system, the Union Carbide Centria®, uses the principle of gel filtration chromatography (see Figure 4). In 1975, Ertingshausen et al.[7] described the Centria® system and the use of Sephadex® G-25 in assays for digoxin, T_3, T_4, and T_3 uptake. As these low molecular weight molecules move through the columns under centrifugal force, the Sephadex® G-25 acts as a molecular sieve, slowing the passage of the unbound antigen while allowing the antigen-antibody complex to pass through unretarded. In 1976, Green et al.[8] described the adaptation of somatotropin (mol wt 21,500) and insulin (mol wt 5,800) radioimmunoassays to the Centria analyzer using disposable plastic columns filled with anionic exchange gel, diethylaminoethyl-Sephadex® A-50. These columns achieve separation by a combination of molecular sieving and ion exchange properties.

The Centria® system is not a fully automated device. Like the RIA-E 6000 system previously described, the operator is required to make manual transfers at two steps during the analysis. The total system comprises three rotating heads which accept disposable transfer discs. In the pipetter module, samples and standards are aspirated, diluted with distilled water, and placed in the cavities of the 36-place transfer disc.

The radioactively labeled antigen and the antibody are physically separated from each other by contoured ridges in the transfer disc. When all reagent transfers are completed, the operator removes the disc and places it on a rotor head in the incubator/separator unit (see Figure 5). Here the disc is surrounded by a ring of collecting tubes in which the Sephadex® gel filtration columns are mounted. When the rotor is started, it accelerates to a low speed, allowing the reagents to be mixed together which simultaneously initiates the incubation. Thus, any question about variation in the assay due to variable incubation time or temperature is eliminated.

At the end of the preset incubation period (typically 5 to 30 min), the rotor accelerates to a higher speed, transferring the mixtures onto the separation columns. An eluant flushes the antibody-bound radioactivity through the columns into the collecting tubes, while leaving the free antigen behind. Once separation is complete, the operator moves the column/tube assembly to the counter/computer module. Here the tubes are counted, three at a time, the raw counts manipulated by an internal microprocessor, and the results printed out in appropriate concentration units.

Recently a second antibody solid phase separation technique has been developed for Centria® by Union Carbide's Bio Systems Division in France. The exchange gel is replaced by a tablet of sephorose-bound antirabbit gamma globulin. Following the initial binding reaction, the reactants are incubated for 5 min with the second antibody. After this, a high-speed wash moves the free antigen into the collecting tube, leaving the antibody-bound complexes in the column. Union Carbide is currently developing assays using this technique for TSH, HPL, ferritin, B-12, folate, and cortisol.

The Centria® system offers the user a reasonable amount of flexibility in setting up different assays. Except for the disposable columns, he may purchase materials from any source or develop his own. Off-line incubation can be accommodated by stopping the disc once the reagents have been mixed together. The radioactive counter allows counting of iodine-125 or cobalt-57; and the microprocessor, an Intel 8080, allows the user a number of options in analyzing the data.

A second system, which uses an anionic exchange column for separation, has been described by Brooker et al.[9] This instrument approach, called Gammaflo®, is commercially available from E. R. Squibb & Sons, Inc. (see Figure 6). The system is based on the bubble-separated continuous flow concept pioneered by Skeggs.[10] Patient samples placed in a 175-sample carrier are picked up by means of a sampler and mixed with labeled ligand and antiserum solution. To reduce the cost per test, the valuable reactants are introduced coincident with each sample and are not wasted between sample draws. Incubation takes place in a timed-delay coil which can be held at constant temperature. The reaction mixture passes out of the delay coil and onto the surface of a mixed-bed separating column containing a separating material chosen for the assay. The thyroxin assay, for example, uses an anion exchange resin which absorbs the free labeled material and allows the complexed weakly charged molecules to pass freely through. Detection of the radioactive mixture triggers a high rate of buffer flow to wash the material through the column and into a sodium iodide counting well. Scaler counting is initiated as soon as the chamber is full, and at the end of a preset time, the chamber is emptied and washed. The counter has been interfaced to an LSI 11 microprocessor which allows the user to choose his own method of data reduction from the many available today.

Unlike other systems in which the operator is involved in transferring materials from one step to the next, the Gammaflo® system is totally automated from sample pickup to the printout of the results. Using this system, Squibb has developed assays for T-4, T-3 uptake, T-3 RIA, digoxin, and cortisol. Other assays currently under development include estriol, angiotensen, progesterone, and testosterone. While the system offers some flexibility in designing one's own protocols and in acquiring materials from any

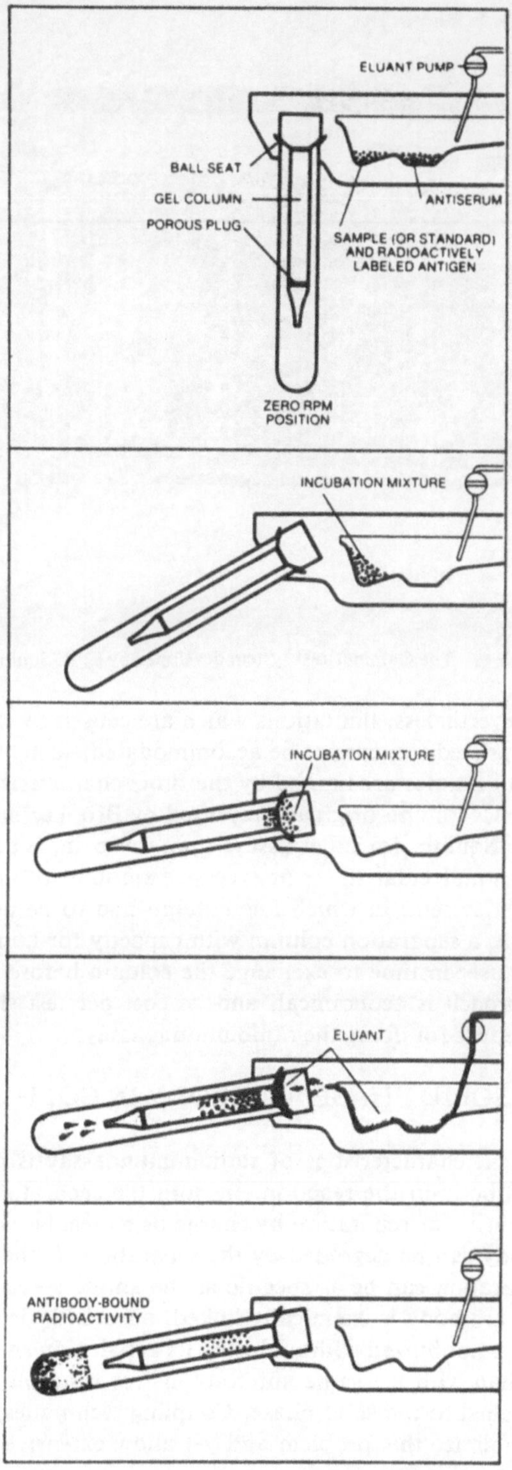

FIGURE 5. Diagrammatic presentation of what happens in the center module of Centria®. Reagents initially separated are mixed and allowed to incubate. They are then spun onto the head of the gel column followed by an eluant for separation.

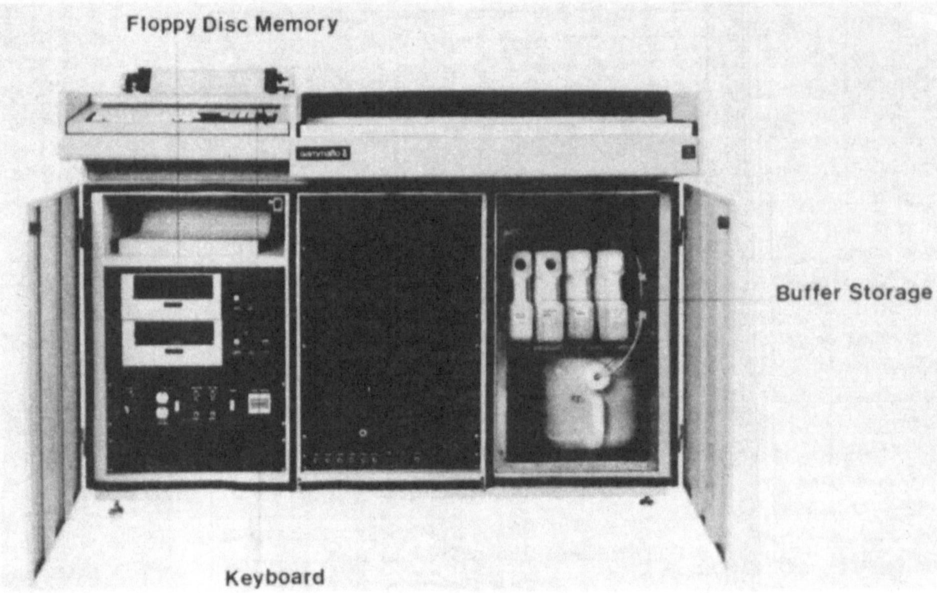

FIGURE 6. The Gammaflo® system developed by E. R. Squibb & Sons.

source, there are, nevertheless, limitations which are caused by the flow-through concept. Certain assay procedures cannot be accommodated, such as the two-site IRMA. In addition, incubation times are limited by the finite characteristics of the delay coil. The charcoal Dowex® column originally reported by Brooker[9] need not limit the separation possibilities. Squibb, for other assays than thyroxin, is developing other separation media, such as molecular sieves or excessive amounts of immobilized antibody. Unlike the Centria® system, in which one column had to be used for each sample, the Gammaflo® uses a separation column with capacity for hundreds of assays. The computer alerts the user in time to exchange the column before its capacity has been saturated. This approach is economical, and its cost per test should be competitive with any other procedure for doing the radioimmunoassay.

V. SOLID PHASE ANTIBODY IN COLUMNS

One of the inherent characteristics of radioimmunoassay is the specificity that is provided by the antibody in the reaction. In both the separation concepts discussed this far, that of filtration or separation by charge or molecular weight, the basic specificity of the antibody can be degraded by the separation. If the antibody is immobilized, however, separation can be as specific as the antibody/carrier combination. In actual practice, the antibody is chemically linked, adsorbed, or physically trapped in such a way that it is immobilized without losing its activity. In some cases, the sensitivity can be poorer than with the same antibody in free solution because a portion of the antibody is attached to the solid phase. Coupling techniques, however, have been developed which minimize this problem and yet allow extensive washing of the solid phase.

The Aria II® automated RIA system, sold by Becton-Dickinson, is a system which uses antibody covalently bound to glass beads packed into a column (see Figure 7). Reese and Johnson[11] have described the measurement of serum thyroxin with the Aria II®. The operator initiates the process by loading a carousel with sample cups containing 200 μl of standards or unknowns. As the cups are indexed one by one, each sample

FIGURE 7. The Aria II® system developed by Becton Dickinson.

is automatically diluted and passed over the immobilized antibody, followed sequentially by a known volume of iodine-125-labeled thyroxin dispensed from an internal supply. The unbound thyroxin, including the free labeled thyroxin, leaves the column and enters a sodium iodide detector where the radioactivity is counted. An eluting agent then passes through the column releasing bound material which is also counted. This process, followed by a wash, regenerates the antibody chamber and readies it for the next sample (see Figure 8). Apparently, the high surface to volume ratio within the antibody chamber allows the immunologic reaction to proceed at maximal rate. The entire cycle, from one sample to the next, for thyroxin, is reported to take only 2.2 min.

Columns containing immobilized antibodies for other molecules can be substituted in Aria II® for the thyroxin column. The system also gives the user the option of using tritium-labeled material by providing a beta scintillator detector and scintillation cocktail reagents within the system. Using the beta emitter, assays have been developed for estriol, progesterone, and cortisol. Estriol and progesterone require extraction of the material from the serum, but all other operations are totally automatic and machine cycle time is only 3 min per test on these assays. A modified logit-log curve, using weighted regression analysis, is generated from the standards. The samples are then calculated against this curve, and final values are printed out in concentration units.

Cottrell et al.[12] have described a system using disposable columns containing immobilized antibodies. The antibody in his system is physically trapped in a polyacrylamide

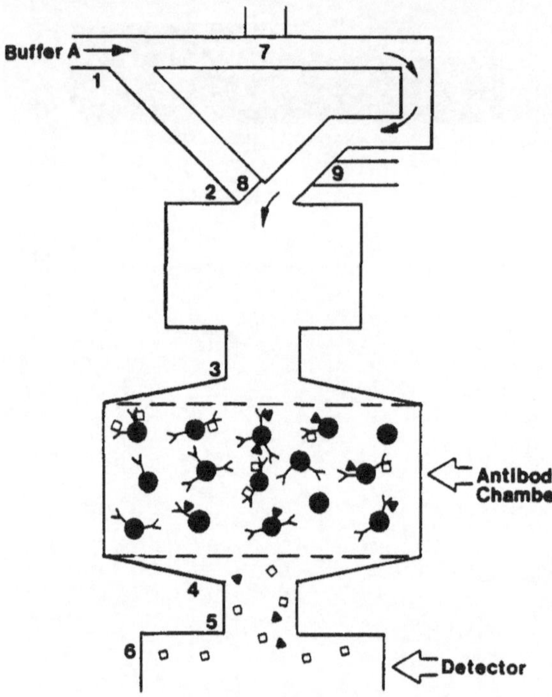

FIGURE 8. Diagrammatic presentation of Aria II® showing the unbound antigen and tag (□, ▲) passing through the column and into the detector. Buffer B subsequently enters via Valve 8, while a rinse solution brought in through Valve 9 and exiting Valve 7 removes residue in the sample loop and aspirator in order to eliminate carryover.

gel with a pore size chosen such that the antibody cannot escape, but the antigens can freely migrate into and out of the gel. This concept offers all of the advantages of immobilized antibody columns without altering the sensitivity of the antibody by chemically linking it to the solid phase. Unfortunately, this approach is not viable for the assays of high molecular weight proteins, for any porosity which would entrap the antibody would exclude the desired protein. This exclusion of high molecular weight molecules, however, can be very beneficial in certain assays where these molecules would interfere with the material to be assayed. In addition, a patented process[24] for preparation of the polyacrylamide gel makes it hygroscopic, such that the reaction materials are drawn into the gel in a volume which is essentially independent of that volume presented. To date, Updike's concept has not been developed into a commercially available system.

VI. ANTIBODY IMMOBILIZED ON MAGNETIC PARTICLES

In 1975, Hersh and Yaverbaum[13] described a radioimmunoassay using a magnetic solid phase. In their assay, digoxin antisera had been coupled to finely divided magnetic oxide particles (Fe_3O_4). The magnetite particles had irregular shapes and diameters which ranged from 1.5 to 10 μm. Separation of the bound and free was accomplished by the use of a U-magnet. A workable standard curve demonstrated for the range 1 to 10 ng digoxin per milliliter plasma. The success of this approach showed that magnetic fields could replace centrifugation in radioimmunoassays and have the potential of simplifying the design of an automated solid phase RIA system.

In November of 1975, Nye et al.[14] published on a solid phase magnetic particle radioimmunoassay in which antibodies were covalently linked to a polymer coated iron oxide (Enzacryl®). This group, at St. Bartholomew's Hospital in London, showed assays for thyroxin, HPL, and digoxin. They advanced the state of the art in equipment design by proposing a magnetic rack which could be used for mixing and separating the magnetic material. The rack could hold 100 tubes in one of two positions, either on the top or suspended between the ridges of the magnets. When the tubes were suspended between the ridges, alternating the magnetic field on and off (2 sec in every 4 sec) provided mixing of the material. Separation of the antibody bound and free fractions was obtained by moving the test tubes directly over the ridges and turning the magnetic field on for approximately 3 min. The magnetic field was held on continuously while supernatant fluid was aspired. The efficiency of this design was demonstrated by the ability to carry out three washes and separations in 15 min for 100 tubes.

Ithakissios et al.,[15,16] working for the Medical Products Division of 3M Company, have advanced the state of the art in the development of magnetic microparticles. Their technology involved glutaraldehyde cross-linking of protein material to form a matrix containing barium ferrite. They demonstrated various formulations in which the protein matrix had specific or nonspecific absorption properties. They found that particle size did not influence the accuracy of the assay, but use of particles less than 100 μm in diameter provided large surface areas and reduced the number of particles needed in the assay. After the particles were formulated with the specific ligand of choice, they were compressed into tablets and subjected to a magnetic field for about 10 sec to permanently magnetize the particles inside before using them in radioassays. This group also described the design of a magnetic rack which allowed handling of 50 tubes simultaneously with good reproducibility (intra assay CV of 2.1%).

The technology of using antibody bound to magnetic particles has been made commercially available in an automated system offered by the Technicon Corporation (see Figure 9). The Technicon system offers complete automation with no operator intervention from sample loading to the printout of results. Because their system uses the continuous flow principles, special requirements were placed on the magnetic particles. They had to be sufficiently magnetic to allow fast, efficient separation, and their physical properties had to be such that they would not block, settle, or break up during the pumping. The Enzacryl® solid phase, developed by Nye and colleagues, was found unsuitable for mixing within the incubation coil due to its bulk density. Magnetic iron oxide (Fe_3O_4) coated with α-cellulose, however, had the desired properties. An optimum particle size was found to be 0.5 to 1 μm. To minimize carry-over, incubation times were held to 10 min. Due to the inherent reproducibility of the continuous flow system, it was found that these incubation times did not result in a loss of sensitivity or precision (see Figure 10). For separation, two electromagnets are used to hold the magnetic particles on the inside of the glass tube while wash solution passes by at a high rate (1 mℓ/min). Magnet 1 is programmed to de-energize before Magnet 2, so that the particles are further washed as they transfer. At the completion of washing, the particles are diverted into a counting coil which maintains them within the sensitive area of a quartz silica 2 in.-diameter sodium iodide detector for at least 80 sec.

A number of assays have been demonstrated on the Technicon automated RIA system. As with some of the other fully automated systems, the applications will be limited by the materials offered by the manufacturer. In the initial marketing, Technicon will have magnetic solid phase antibody and other reagents for T_3, T_4, T_3 uptake, digoxin, cortisol, and HPL. In addition, a "second" antibody solid phase will be available for users who want to develop their own assays using antiserum in solution which gets "picked up" by the second antibody solid phase.

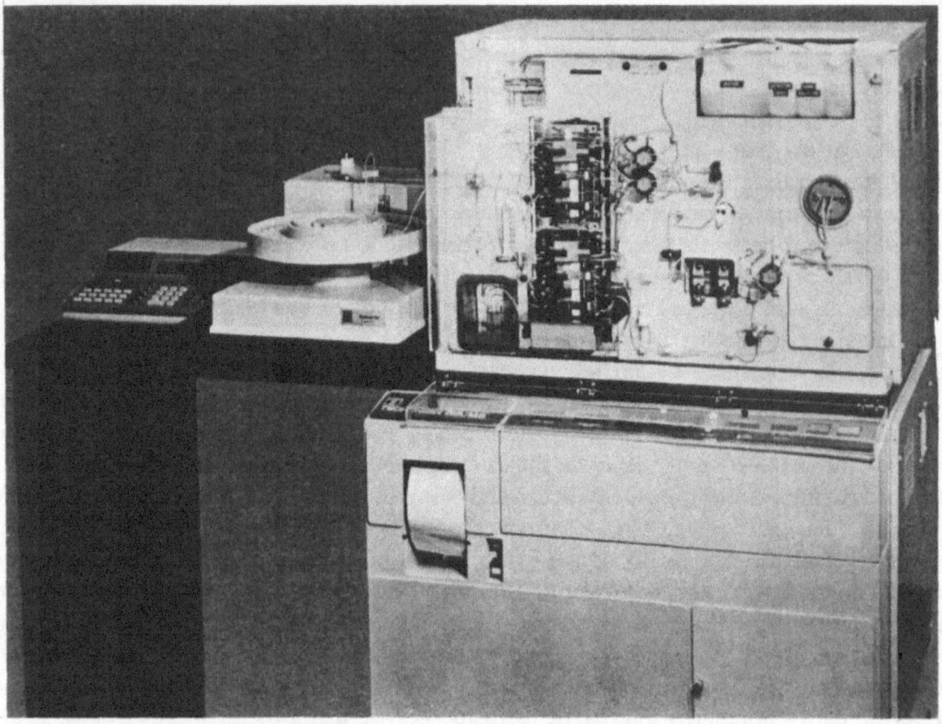

FIGURE 9. The Star system developed by Technicon.

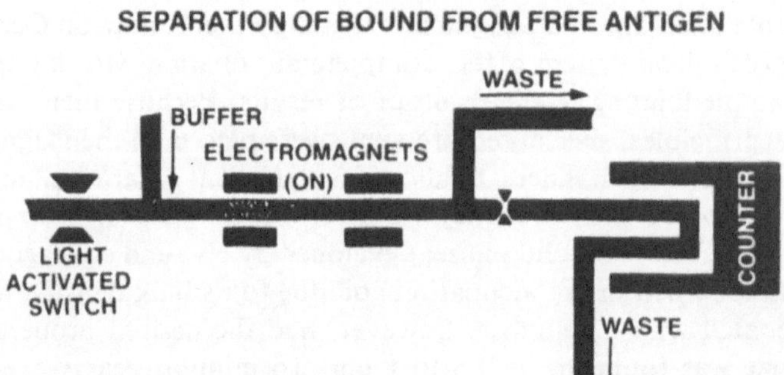

FIGURE 10. Diagrammatic presentation of the separation and counting process used in Star. When the electromagnets are turned on the magnetic solid phase is held; when they are released, the magnetic particles go into the counter for quantitation and then to waste.

VII. ANTIBODY COATED ON PLASTIC TUBES

Another form of immobilizing antibody was reported by Catt and Tregear[17] in 1967. This technique makes use of the physical adsorption of gamma globulin onto polypropylene plastic surfaces, typically the inner wall of test tubes. Use of the tubes is extremely simple. Only two pipetting steps are required (sample and buffer containing radioactive label), and separation of the bound and free fractions is merely a matter of emptying and washing the tubes. The bound fraction stays attached to the tube walls for counting.

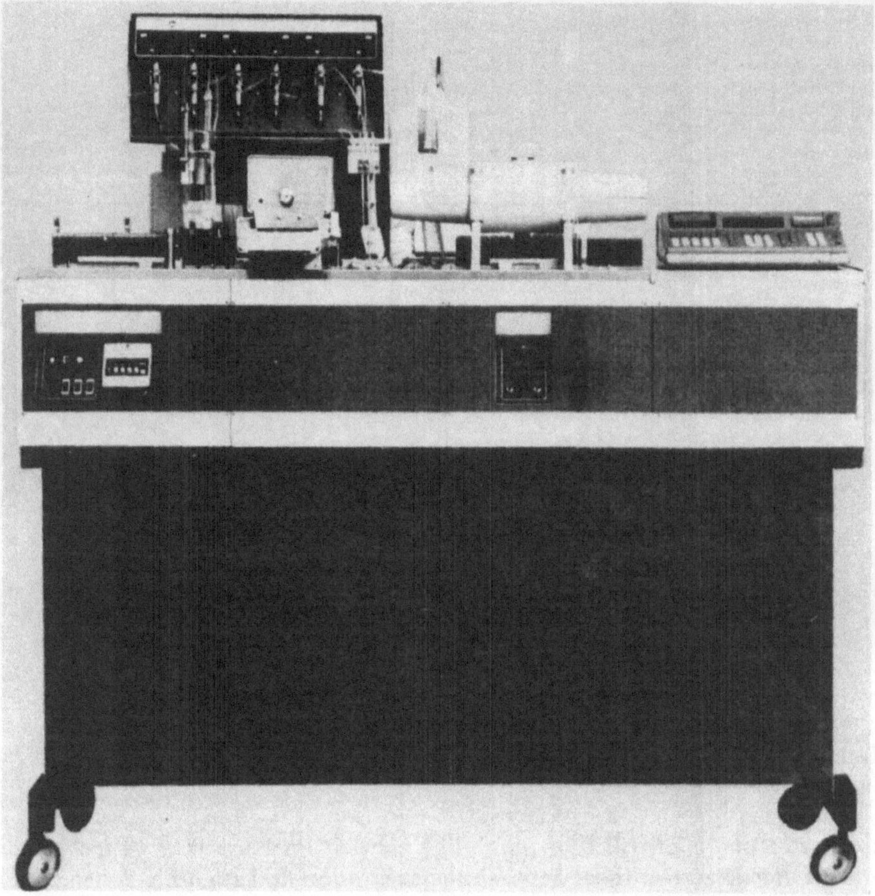

FIGURE 11. The Concept 4 system developed by Micromedic Systems.

Table 1
CONCEPT 4 STANDARDIZED ASSAY PARAMETERS

1. Add sample to antibody-coated tube (2 to 400 μl)
2. Add buffer containing radioactive label (100 to 1000 μl)
3. Vortex
4. Incubate (0 to 999 min at ambient or 37°C)
5. Aspirate and wash twice
6. Count (iodine-125 or cobalt-57)
7. Calculate
8. Print result

The first fully automated immunoassay system commercially available, Micromedic Systems Concept 4® automatic radioassay[18] uses 8 × 50-mm plastic tubes coated with specific antibody for each test performed (see Figure 11). A standard protocol consists of the eight steps in Table 1.

The only manual manipulation required is the loading of samples, buffer, and coated tubes into the system. Each tube moves through the instrument in a six-second cadence, so that each receives identical treatment in mixing, incubation, and separation (see Figure 12). The counter has dual crystals for counting two tubes simultaneously,

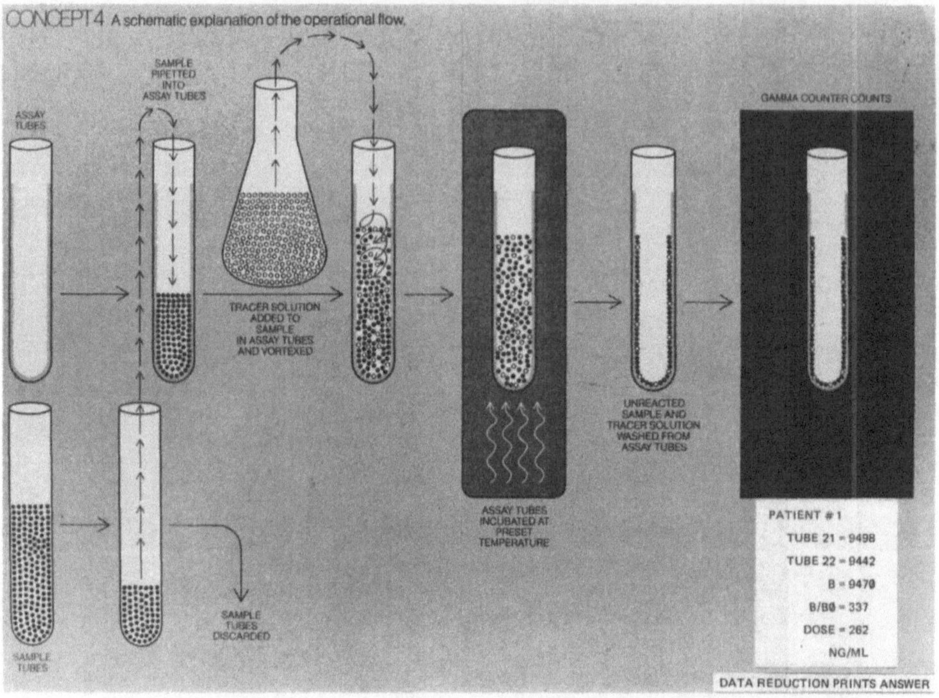

FIGURE 12. Diagrammatic presentation of the coated-tube approach used for separation in the Concept 4.

and data reduction is provided by an on-line Hewlett Packard 9815. A rack accumulation zone for up to 600 tubes between the separation and counting systems accommodates the different throughput rates of the liquid handling and gamma-counting operations. One of the primary advantages of this system over others available is that in the event of an instrument breakdown the entire assay can be run manually.

A system described by Miles,[19] but not commercially available, uses antibody coated on the inner walls of standard polypropylene pipet tips. A specially modified geometry using polystyrene, described by Park,[20] works on this same principle. The ease with which reagents can be drawn in and expelled from these tips make a number of protocols possible which would be tedious or impossible in other systems. For example, it is possible to take repeated aliquots and do extraction or concentration of the unknown or multiple sequential reactions, such as the immunoradiometric assay (IRMA) pioneered by Miles and the two-site IRMA or "sandwich" assay.[21]

VIII. HOMOGENEOUS ASSAYS

The ultimate simplicity in separating bound from free is achieved in so-called homogeneous assays for which a physical separation is not required at all. Such a result is achieved by using as a label, not a radioactive isotope, but an enzyme which is either inhibited or reactivated when the antigen to which it is attached complexes with an antibody. This approach, called EMIT (enzyme-multiplied immunoassay technique), has been well established by SYVA Corporation for therapeutic anticonvulsant drug monitoring and was recently made available for T-4 and digoxin. In the case of the T-4 assay, for example, malate dehydrogenase activity is greatly reduced when it is chemically bound to thyroxine. When the T-4 enzyme complex subsequently binds to antibody specific for T-4, the enzyme is reactivated. Hence, in a competitive assay with human serum, there is an inverse relationship between the amount of T-4 in the patient sample and the enzyme activity. This makes it possible to automate the T-4 assay using standard enzyme methodologies. Successful automation using this approach has been reported for the Auto Chemist instrument[22] and the Abbott ABA-100[23] microchemistry analyzer. One of the main advantages of this technique is the elimination of radioactive isotopes. For a high-volume assay such as T-4, this can greatly simplify laboratory procedures because reagent shelf life is greater, radiation monitoring is eliminated, and specialized equipment for counting is avoided.

One of the major drawbacks of the EMIT technology is that the lower limit of detection is probaly about 1 $\mu g/d\ell$. This will limit the range of tests which can make use of this technique. Other nonisotopic immunoassay techniques currently under development have potential for greater sensitivity and will certainly play a role in the future automation of these procedures.

IX. CONCLUSIONS

In the absence of the automated equipment described here, there are still numerous products which compete in the RIA hardware market. Current laboratory practice makes use of:

- Semiauto or automatic pipetter/diluters
- Gamma counters
- Data reduction computers
- Systematized components

Increasing test volumes and demand for new tests will cause laboratory directors to consider purchase of additional equipment. Automated equipment such as we have described here might be purchased to:

- Reduce the number or skill level of technologists required to perform the routine assays
- Reduce the test cost
- Reduce the turn-around time
- Improve the accuracy or precision of results
- Accommodate expanded test repertoire
- Reduce the pressure on existing equipment by handling all the high-volume routine testing
- Upgrade worn-out, obsolete, or inadequate equipment

Some of the commercially available systems which we have described are summarized in Table 2 which can be used for initial comparison to lab requirements. Whether to purchase automatic equipment or not and which system suits the lab best must be evaluated carefully by each laboratory. Individual turn-around time commitments, planned test repertoires, test volume, labor costs, and expectations of performance parameters can make one system more appropriate than another for a given lab. We are fortunate that technology has offered so many alternatives.

Table 2
COMPARISON OF AUTOMATED RIA SYSTEMS

Instrument	RIA-E 6000	Centria	Concept-4	Aria II	Gammaflo	Star	Pace-4
Manufacturer	Laboratorium, Professor Dr. Berthold, D-7547 Wildbad 1, West Germany	Union Carbide Corp., 401 Theodore Fremd Ave., Rye, N.Y., 10580	Micromedic Systems, 102 Witmer Rd., Horsham, Pa., 19044	Becton Dickinson, Mountain View Ave., Orangeburg, N.Y., 10962	E. R. Squibb & Sons, Inc., P. O. Box 4000, Princeton, N.J., 08540	Technicon Instruments Tarrytown, N.Y., 10591	Picker Corporation, 12 Clintonville Rd., Northford, Conn., 06472
Type of system	Semiautomatic, discrete, sequential	Semiautomatic, discrete, simultaneous	Automatic, discrete, sequential	Automatic, continuous flow, sequential	Automatic, continuous flow, sequential	Automatic, continuous flow, sequential	Automatic, discrete, sequential
Date available	Fourth quarter 1976	First quarter 1977	First quarter 1977	Second quarter 1977	First quarter 1979	Second quarter 1979	First quarter 1981
System price ($)	65,000	39,000	65,000	59,500	55,000	35,000	70,000
Reagent price ($)	Standard reagents + 0.20 per filter	0.55—0.80 per analysis	0.50—0.75 per analysis	0.35—0.50 per anaylsis	Average 0.50 per analysis	0.70—1.00 per analysis	0.40—0.65 per analysis
Separation technique	Discrete filters on tape separate any insoluble complex	Individual gel filtration chromatography columns	Antibody-coated 8 × 50 mm test tubes	Reusable immobilized antibody column	Reusable anion-exchange column or other material	Immobilized antibody on magnetic particles	Discrete filters on tape
Initial loading capacity	650 tubes	36 tubes per disc	200 tubes	120 tubes	175 tubes	40 tubes	160 tubes
Time to get first output	Depends on protocol	30—45 min	0—999 min (typical 15)	3 min	8—15 min	15 min	Depends on protocol
Throughput rate per hour	220 tubes	72 tubes	120 tubes	20—30 tubes	40—60 tubes	60 tubes	220 tubes
Assays available	Numerous assays have been developed by users	T_4, T_3 RIA, T_3 uptake, digoxin, insulin	T_4, T_3 RIA, T_3 uptake, digoxin, cortisol, insulin, TSH	T_4, estriol, progesterone, cortisol, T_3 uptake, digoxin	T_4, T_3 RIA, T_3 uptake, digoxin, cortisol	T_4, T_3 RIA, T_3 uptake, digoxin, cortisol, HPL	Under development; see text section entitled "Filtration Techniques"

REFERENCES

1. **Berson, S. A. and Yalow, R. S.,** General principles of radioimmunoassay, *Clin. Chim. Acta,* 22, 51, 1968.
2. **Skelley, D. S., Brown, L. P., and Besch, P. K.,** Radioimmunoassay, *Clin. Chem.,* 19, 146, 1973.
3. **Peron, F. G. and Caldwell, B. V.,** *Immunologic Methods in Steroid Determination,* Appleton-Century-Crofts, New York, 1970.
4. **Walker, W. H. C.,** An approach to immunoassay, *Clin. Chem.,* 23, 384, 1977.
5. **Marschner, I., Erhardt, F., Henner, J., and Scriba, P. C.,** A modular analyzer system for double antibody radioimmunoassays, *Z. Klin. Chem. Klin Biochem.,* 13, 481, 1975.
6. **Bagshawe, K. D.,** Computer controlled automated radioimmunoassay, *Lab. Pract.,* 24(9), 573, 1975.
7. **Ertingshausen, G., Shapiro, S. I., Green, G., and Zborowski, G.,** Adaptation of a T_3-uptake test and of radioimmunoassays for serum digoxin, thyroxine, and triiodothyronine to an automated radioimmunoassay system-"Centria", *Clin. Chem.,* 21, 1305, 1975.
8. **Green, G., Dyce, D., Gimovsky, A., and Lo, D. H.,** Automated approach to radioimmunoassays of somatotropin (human growth hormone) and insulin, *Clin. Chem.,* 22(9), 1510, 1976.
9. **Brooker, G., Terasaki, W. L., and Price, M. G.,** Gammaflow: a completely automated radioimmunoassay system, *Science,* 194, 270, 1976.
10. **Skeggs, L. T.,** An automatic method-colorimetric analysis, *Am. J. Clin. Pathol.,* 28, 311, 1957.
11. **Reese, M. G. and Johnson, L. R.,** Automated measurement of serum thyroxine with the "Aria II" as compared with competitive protein binding and radioimmunoassay, *Clin. Chem.,* 24(2), 342, 1978.
12. **Cottrell, David B., Toren, E. C., Jr., Magnusen, J. A., and Updike, S. J.,** Computer controlled automation of radioimmunoassay based on gel entrapment of antibody, *Clin. Chem.,* 20(7), 829, 1975.
13. **Hersh, L. S. and Yaverbaum, S.,** Magnetic solid-phase radioimmunoassay, *Clin. Chim. Acta,* 63, 69, 1975.
14. **Nye, L., Forrest, G. C., Greenwood, H., Gardner, J. S., Roberts, J. R., and Landon, J.,** Solid-phase, magnetic particle radioimmunoassay, *Clin. Chim. Acta,* 69, 387, 1976.
15. **Ithakissios, D. S. and Kubiatowicz, D. O.,** Use of protein containing magnetic microparticles in radioassays, *Clin. Chem.,* 23(11), 2072, 1977.
16. **Ithakissios, D. S., Kubiatowicz, D. O., Windorski, D. C., and Wicks, J. H.,** Immune and non-immune T_4 radioassays utilizing albumin magnetic microparticles, *Clin. Chim. Acta,* 84, 69, 1978.
17. **Catt, K. J. and Tregear, G. W.,** Solid-phase radioimmunoassay and antibody-coated tubes, *Science,* 158, 1570, 1967.
18. **Rugg, N. E., Hasler, M. J., Bjornsen, R. E., and Painter, K.,** A fully automated solid-phase radioimmunoassay for triiodothyronine, *Clin. Chem.,* 23(5), 851, 1977.
19. **Miles, L. E. M.,** U.S. Patent 4,087,248, 1978.
20. **Park, H.,** A new plastic receptacle for solid phase immunoassays, *J. Immunol. Methods,* 20, 349, 1978.
21. **Miles, L. E. M.,** Properties, variants, and applications of the immunoradiometric assay method, *LaRicerca Clin. Lab,* 5(1), 59, 1975.
22. **O'Brien, J. E.,** Automated T_4 testing by EMIT, *Lab. Manage.,* 16, 50, 1978.
23. Syva, Syva Clinical Trial Report #37 Palo Alto, Calif., 1977.
24. **Updike, Stuart J.,** U.S. Patent 3,925,017, 1975.

INDEX

A

Absorbance, 4
Absorptiometry, see also Densitometry, 5
Acebutolol, 89
Acedapsone, 90
Acetaminophen, 88
N-Acetylprocainamide, 89
  HPLC, 92
Adenosine, 82
Adsorbtion chromatography, 73
  methyl substituted purines, 74
Albumin, 83
Alkaloid, see also specific compounds
  HPLC, 84
Allopurinol, 88
ALU, see Arithmeticlogic unit
Amino acid, see also specific compounds
  HPLC, 81—82
  analyzer, 79
p-Aminobenzoic acid (PABA), 82, 84
ε-Aminocaproic acid, 81—82
Amitryptyline, 87
  HPLC, 90
Amobarbital, 85
Amperometric detection, 78
Analgesic, see also specific compounds
  HPLC, 88
Analog to digital converter, 65
Anesthetic, see also specific compounds
  HPLC, 89
Antiarrythmic drug, see also specific compounds
  HPLC, 89
Antibiotic, see also specific compounds
  HPLC, 89—91
Antibody, immobilized, see Radioimmunoassay, immobilized antibody
Anticonvulsant, see also specific anticonvulsants
  HPLC, 85—86
Antidepressant, see also specific compounds
  HPLC, 86—87
Antimicrobial susceptibility, 108—109
Aprobarbitone, 85
Aria II®, automated RIA system, 136—137, 144
  diagrammatic presentation, 138
Arithmetic-logic unit (ALU), 59
ATP, microbial detection, 117—118
Autism, HPLC analysis, 76
Autobac 1, automated microbiology system, 114
Automated leukocyte differential analyzer, see Leukocyte differential analyzer
Automated microbiology, see Microbiology
Automated radioimmunoassay, see Radioimmunoassay
Auto-Microbic System® (AMS®), automated microbiology system, 112—113

B

Bactec®, automated microbiology system, 110—111
Bactometer®, automated microbiology system, 111—112
Balance, microprocessor-based, 69
Barbitone, 85
Barbiturate, see also specific compounds
  HPLC, 85
Basophil, 37
Bauer-Kirby method, see Disk diffusion
Bayes' theorem, 106—107
Benzodiazepine, see also specific compounds
  HPLC, 85
Benzoylecognine, 89
Beer-Lambert law, 4
Binary number system, 59
Bit, 59
Blood gas analyzer, 64
Bonded phase support, 72
Byte, 59

C

Cadaverine, 83
Caffeine, 74—75, 87
Cannabis, 84
Carbamazepine, 85—86
Carboxylic acid, see also specific compounds
  HPLC, 82
Catecholamine, see also specific compounds
  HPLC, 82
Cefoxitin, 90
Centria®, automated RIA system, 133—134, 144
  mixing and incubation, 135
Cephalosporin, 90
Cephalothin, 90
β-Cetotetrine, 89
Chlordiazepoxide, 85
Chlorimipramine, 87
  HPLC, 89
p-Chlorophenoxyisobutyric acid, 91
Chlorpromazine, 87
Chlorpropamide, 91
  HPLC, 93
Chromatography, see specific types
Clofibrate, 93
Clofibrinic acid, 91
Cocaine, 89
Computer, digital, see Digital computer
Concept 4®, automated RIA system, 141, 144
  diagrammatic presentation, 142
Connectivity, 44
  algorithm, 45—46
Control subsystem, 59
Cortisol, 80
Cotinine, 85
Creatinine, 82
Cyclobarbital, 85

D

Daunomycin, 90
Daunomycinol, 90
Decimal system, 58
Densitometer, scanning, 2—4
  accuracy
    auto-gain, 27, 32
    integration, 26—27, 32
    pen recorder, 25—26, 32
    zeroing, 27, 32
  linearity, 25, 31—32
    colored dyes, 30—31
    neutral density sample, 29—30
  optical slit and off-image blocking, 27—28
  range, 24—25
  rise time, 28—29
  scan rate, 26, 32
  slit response profile, 32—33
Densitometry, 2—4
  bandwith error, 13—16
  dual beam, 6, 12
  fixed offset errors, 12—13
  fluorometer/densitometer, see Fluorometer/densitometer
  instrumentation, see also Densitometer, 5—6
  light-source/detector stability, 6—7
  slit size, 7—12
  theory, 4—5
  wavelength selection, 16
N-Desmethyldiazepam, 85
Desmethylimipramine, 86—87
Detector, see specific types
Diaminodiphenylsulfone (DDS), 90
Diazepam, 85
  HPLC, 88
Differential III®, automated microbiology system, 114
Differential light scattering (DLS), 113—114
Diffuse density, 5—6
Diffalone, 93
Digital computer, 69—70
  applications, 63
    calibration, 67
    data input, 64—65
    derived results, 67—68
    limitation of, 68
    measurement techniques, 65—67
    sequencing and control, 64
    simplification, 68
  classes of, 61, 63
  future developments, 68—69
  history of, 58
  number systems, 58—59
  operation, 60, 62—63
  subsystems, 59—61
Digital to analog converter, 65
Digoxin, 138
Dihydromorphine, 84
1,25-Dihydrovitamin D, 80
3,4-Dihydroxyphenylacetic acid, 82
Diisopyramide, 89
Diphenylhydantoin, 85—86
Disk diffusion, 108—109
L-Dopa, 82
  HPLC, 78
Dopamine, 82
  HPLC, 78
Double-antibody method, 129, 132
Dual slope interating converter, 65
Dyphylline, 87

E

Electrochemical detection, 78
Electrophoresis
  automated microbiology, 120
  quanitation
    densitometric, see Densitometry
    fluorometric, see Fluorometry
EMIT, 142—143
Enzyme
  multiplied immunoassay technique, see EMIT
  profile, 118—119
Eosinophil, 37
Ergotamine, 85
Erythrocyte, 37
Estradiol, 80
Estriol, 80
Estrone, 80
Ethchlorvynol, 86
Ethosuximide, 86
Excited state, 16
Extinction coefficient, 5

F

FDA, see Food and Drug Administration
Flunitrazepam, 85
Fluorometer, scanning, 2—3
  auto-gain accuracy, 27, 32
  integration accuracy, 26—27, 32
  linearity, 25, 31—32
  optical slit and off-image blocking, 27—28
  pen recorder accuracy, 25—26, 32
  range, 24—25
  rise time, 28—29
  scan rate, 26—32
  slit response profile, 32—33
  zeroing accuracy, 27, 32
Fluorometer/densitometer, scanning, 2—3
  Corning 720®, 21—22
    electronics, 24
    optical system, 23
Fluorometry, 2—4
  automated microbiology, 119—120
  detector, 76—78
  errors, 20—21
  instrument for, see Fluorometer
  theory, 16—18
    fluorescence activity, 18—20

Fluphenazine, 87
Food and Drug Administration (FDA), 121
Ftorafur, 90

G

Gammaflo®, automated RIA system, 134—136, 144
Gas chromatography, 72
automated microbiology, 118—119
computer controled, 66
Gel filtration chromatography, see Gel permeation chromatography
Gel permeation chromatography, 73
radioimmunoassay, 133
Glycosphingolipid, see also specific compounds
HPLC, 80—81
Golay's algorithm, 47—49
Gout, 88
Griscofluvin, 91
Ground state, 16

H

Hemalog-D®, leukocyte differential analyzer, 38
Hematrak®, leukocyte differential analyzer, 38
Hemodialysate, 79
Hemoglobin, 83
HPLC, 86
Heptabarbital, 85
Heptabarbitone, 85
Heroin, 84
Hexobarbital, 85
HPLC, 87
High-performance liquid chromatography (HPLC), 72—73
alkaloid, 84
amino acid, 81—82
analgesic, 88
anesthetic, 89
antiarrythmic drug, 89
antibiotic, 89—91
antidepressant, 86—87
carboxylic acid, 82, 84
catecholamines, 82—83
column selection, 73—76
detection, 76—79
lipids, 80—81
metabolic profile, 79—80
nucleotides, 82
polyamines, 83
porphyrin, 83
protein, 83
stimulant, 85
theophylline, see Theophylline, HPLC
tranquilizer, 86—87
Histone, 83
Homovanillic acid, 82
HPLC, see High-performance liquid chromatography

Hydrochlorthiazide, 93
Hydroxynalidixic acid, 90
5-(4-Hydroxyphenyl)-5-phenylhydantoin, 86

I

Imipramine, 86—87
Indole compound, see also specific compounds, 82
Indomethacin, 88
Insulin, 83
Iodine-125, 128
Ion-exchange chromatography, 73—74
methyl substituted purine, 75
Ion-pair partition chromatography, 72

K

Ketamine, 89
α-Keto glutaric acid, 82
Kirby-Bauer method, see Disk diffusion

L

LARC®, leukocyte differential analyzer, 38
cell search pattern, 42
Leukocyte, see also specific types
classification, 37
differential count, see also Leukocyte differential analyzer, 36—37
Leukocyte differential analyzer, 36
accuracy, 54—55
automation, 39—40
hardware subsystem, 40—44
block diagram, 41
histochemical recognition, 39
history, 37—39
pattern recognition, 39, 50—53
feature extraction, 50
image processing, 44—50
sample preparation, 53—54
Lidocaine, 89
Linearity factor (L), 14—15
Linear sensor, 67
Lipid, see also specific lipids
HPLC, 80—81
Liquid chromatography, high performance, see High performance liquid chromatography
LSD, 84
HPLC, 77
Luciferase, 117
Luciferin, 117
Lymphocyte, 37

M

Marijuana, see Cannabis

Medical Device Amendment of 1976, 121—122
Melatonin, 82
Methadone, 84
Methimazole, 93
Methsuximide, 86
MIC, see Minimum inhibitory concentration
Microbiology, automated, 104—105
  antimicrobial susceptibility, 108—109
  detection, 105—106
  future projections, 122
  identification, 106—107
  laboratory considerations, 120—122
  methods
    ATP detection, 117—118
    electrophoresis, 120
    enzymatic profile, 118—119
    fluorescent spectroscopy, 119—120
    gas chromatography, 118—119
    impedance, 111—112
    light scattering, 113—115
    microcalorimetry, 116—117
    optical density, 115—116
    particle analysis, 120
    plate streaking, 116
    radiometric detection, 109—111
    selective growth, 112—113
Microcalorimetry, 116—117
Microcomputer, 63
  single-chip, 64
Microparticulate support, 72
Microprocessor, 63
Midicomputer, 63
Minicomputer, 63
Minimum inhibitory concentration (MIC), 108—109
Morphine, 84
MS-2®, automated microbiology system, 115
  bacterial growth monitored by, 116

N

Nalidixic acid, 90
Neisseria gonnorrhoeae, 106
Neutrophil
  band, 37, 47—48
  digitized image, 45
  optical density distribution histogram, 47
  segmented, 37, 47—48
Nicotine, 85
Nicotinic acid, 80
Nitrazepam, 85
Normalization, 23
Nortriptyline, 87
Nucleoli location and identification, 49
Nucleoside, 82
Nucleotide, see also specific nucleotides
  HPLC, 82
Number system, 58—60

O

Optical density, 5
Oxipurinol, 88

P

Pace-4, automated RIA system, 130—132, 144
Paraquot, 84
Parkinson's disease, 82—83
Partition chromatography
  ion pair, 72, 74—76
  liquid-liquid, 74
Pellicular support, 72
Pemoline, 85
Perphenazine, 87
pH electrode, 64—65
  response, 66
Phenacetin, 88
Phenobarbital, 85—87
Phenylpyruvic acid, 82
Phenytoin, 86
Phosphadityl choline, 80—81
Phosphorescence, 16
Pill counter, computer controlled, 60, 62—63
Polyamine, see also specific compounds
  HPLC, 83
Porphyrin, see also specific compounds
  HPLC, 83
Prednisolone, 80
Primidone, 86
Procainamide, 89
  HPLC, 92
Protein, see also specific compounds
  HPLC, 83
Purine, 82
Putrescine, 83
Pyruvic acid, 82

Q

Quantitation efficiency (QE), 8
  amplitude, 10—12
  area, 9—10, 12
Quantum efficiency, 16
Quinidine, 89

R

Radioimmunoassay (RIA), 128
  automation of, 128
    systems comparison, 144
  conclusions, 143
  filtration, 129—132
  homogenous assay, 142—143
  immobilized antibody
    coated plastic tubes, 140—142
    magnetic particles, 138—140
    solid phase in columns, 136—138
  separation by charge or molecular weight, 133—136

RAM, see Random access memory
Random access memory (RAM), 59
Read only memory (ROM), 59
Refractive index detector, 76
Reversed phase chromatography, 73
  resolution of methyl substituted purines, 75
RIA, see Radioimmunoassay
RIA-E6000, automated radioimmunoassay system, 129—131, 144
  filtration area, 131
Riboflavin, 80
Rise time, 28
ROM, see Read only memory

S

Salicylic acid, 88
Salicyluric acid, 88
Sayce pattern, 33
Scan-rate uniformity, 26
Scene segmentation, 45—47
Secobarbital, 85
Sedative-hypnotic drug, see also specific compounds
  HPLC, 85
Semispecular density, 5—6
Sensor, see specific types
Soap chromatography, 82—83
Spectral bandpass error, 13—16
Specular density, 5—6
Spermidine, 83
Sphingomyelin, 80
Star system, automated radioimmunoassay, 139—140, 144
Steroid, see also specific steroids
  HPLC, 80
Stimulant, see also specific compounds
  HPLC, 84
Successive approximation converter, 65
Sulfinpyrazone, 88
  HPLC, 91
Sulfone, see also specific compounds
  HPLC, 90—91

T

T-4, 142—143
Technicon Auto Analyzer®, 79
Theobromine, 74—75, 87
Theophylline, 74—75
  HPLC, 87—88
Thioridazine, 87
Thrombocyte, 37
Thyroxin, 134, 137
Tolbutamide, 91
  HPLC, 93
Tranquilizer, see also specific compounds
  HPLC, 86—87
Transmission, 5
Transmittance, 4

U

UV-visible photometer, 76

V

Vanilmandelic acid, 82
Vitamin, see also specific vitamins
  HPLC, 80
Vitamin A, 80

W

Warfarin, 84

X

Xanthine, see also specific compounds
  HPLC, 87—88